Counting the Cost of Global Warming

Counting the Cost of
GLOBAL WARMING

John Broome

A report to the
Economic and Social Research Council
on research by John Broome and David Ulph

The White Horse Press

First published 1992

The White Horse Press
10 High Street, Knapwell, Cambridge CB3 8NR, UK
1 Strond, Isle of Harris, Western Isles PA83 3UD, UK

British Library Cataloguing in Publication Data

Broome, John
    Counting the cost of global warming
I. Title
551.6

ISBN: 1–874267–01–4
ISBN: 1–874267–00–6 (pbk)

Typeset by John Broome
Photosetting by the University of Bristol Printing Unit
Printed in Great Britain by the Redwood Press Ltd., Melksham

Contents

Contents

Preface

At the beginning of its Global Environmental Change Programme, the Economic and Social Research Council commissioned a number of 'desk studies' on aspects of climate change. David Ulph and I were asked to investigate the intergenerational aspects of climate change, under an ESRC research grant, number W115 25 1001. This is the report that resulted. I am extremely grateful to David Ulph for his major contributions to the work.

The ESRC's desk studies were intended to help set the programme for future research. This report sets out the issues that are raised by global warming in the area of relations between generations, and suggests directions where research is needed. It is intended to prepare the way by mapping the territory. This is an area where the work of philosophers and economists overlaps. Generally, each discipline has simply ignored the other, and when there has been some communication there has also often been some misunderstanding. This report tries to bring together work in the two disciplines. It is chiefly intended for economists, but I hope other social scientists and philosophers will also find it useful. Sections 3.2 and 3.3 deal with issues that are the particular concern of economists, and other readers may safely skip them. Elsewhere in the report I have assumed

vii

only a slight acquaintance with economics.

This report has many faults. No doubt many result from my own failings, but some also result from the pressure of time. To be useful, the ESRC's desk studies needed to be completed quickly. This report is a quick guide to the issues, and not a compendium. I do not even pretend to have presented a thorough survey of the literature; with a literature ranging from meteorology to ethics, I could not have done that in any amount of time. I have laid out the arguments as I understand them, and drawn what conclusions I can. I have offered definite views on many points, and I have made definite suggestions about what needs to be done next. But I have also left many questions open because I did not have the time to pursue the arguments to the end.

I am grateful to all the following people, who either supplied me with copies of their own writings, or guided me to the writings of others: Wilfred Beckerman, David Braybrooke, Ann Broome, Tyler Cowen, Angus Deaton, Jonathan Fisher, Peter Laslett, Dale Jamieson, John Pezzey, Larry Temkin and David Thomson. I am particularly grateful to Partha Dasgupta, David Donaldson, Derek Parfit, David Pearce, Michael Spackman and Hans-Peter Weikard for the comments I received from them on an earlier draft of the report. They have saved me from numerous errors. I must also thank Andrew Johnson for his very careful work on the proofs, and for writing the index.

Chapter 1
Predictions

Global warming raises unique questions about our responsibilities to future generations. This chapter reviews the
present state of scientific opinion on the future of the earth's
climate, and draws some conclusions about the nature of the
issues we are faced with. It is intended to suggest what
questions future research needs to concentrate on. Later
chapters in this study develop some of the suggestions in
more detail.

My summary of scientific opinion is contained in Sections
1.1 and 1.2. Section 1.3 considers the likely effects of global
warming on human life, and Section 1.4 draws conclusions
aimed at determining the direction of the work in the rest
of this study.

In 1988, the World Meteorological Organization and the
United Nations Environment Programme together established the Intergovernmental Panel on Climate Change
(IPCC). This Panel in turn convened three working groups.
The first was a group of scientists charged with assessing
the available scientific evidence on climate change. The
conclusions of this scientific working group were published
in 1990,[1] and so too was a useful *'Policymakers' Summary'*
of them.[2] The IPCC's scientific assessment has an impec-
cable authority. It represents a very high degree of consen-

sus amongst the scientific community: 'most of the active scientists working in the field' were involved in preparing this report.[3] Since I am not a scientist, I am not well qualified to judge the reliability of other sources. The predictions contained in Sections 1.1 and 1.2 are therefore drawn exclusively from the IPCC's *Policymakers' Summary*. In these sections, comments and interpretations drawn from other sources are confined to notes.

1.1 *Greenhouse gases*

Greenhouse gases are more transparent to the short-wavelength radiation emitted by the sun than they are to the longer-wavelength radiation emitted by the surface of the earth. Consequently, greenhouse gases in the atmosphere tend to warm the earth by trapping radiation.

The most abundant greenhouse gas is water vapour. But changes in the weather cause the atmospheric concentration of water vapour to vary widely and rapidly. Water vapour is therefore always treated endogenously in models of the atmosphere. The other important greenhouse gases are, in order of their present warming effect, carbon dioxide, methane, chlorofluorocarbons and nitrous oxide. (Ozone may also be important, but its effect is not yet quantified.) These gases have various sources and sinks, many of which are influenced by human activity. At present the concentrations of all of them in the atmosphere are increasing. These are the greenhouse gases I shall be concerned with.

Projections for their future concentrations depend, naturally, on the quantities that will be released by human activity. The IPCC set up four possible 'emissions scenarios', and investigated the consequences of each. One, called the 'business-as-usual scenario', is described as follows:

> Population was assumed to approach 10.5 billion in the second half of the next century. Economic growth was assumed to be 2–3% annually in the coming decade in the

OECD countries and 3–5% in the eastern European and developing countries. The economic growth levels were assumed to decrease thereafter. . . The energy supply is coal intensive and on the demand side only modest efficiency increases are achieved. Carbon dioxide controls are modest, deforestation continues until the tropical forests are depleted and agricultural emissions of methane and nitrous oxide are uncontrolled. For CFCs the Montreal Protocol is implemented albeit with only partial participation.[4]

The other scenarios maintain the same assumptions about population and economic growth, but assume more stringent measures are taken to control the release of greenhouse gases. In 'scenario D', for instance, there is a shift to renewable and nuclear sources of energy in the first half of the next century, together with other tight controls on emissions.

Before the industrial revolution, the concentration of carbon dioxide in the atmosphere was 280 ppmv (parts per million by volume). It is now 350 ppmv. On the business-as-usual scenario it is expected to reach 560 ppmv – double its pre-industrial level – by about 2055. By 2100 it is expected to be 830 ppmv.

Concentrations of other greenhouse gases have also increased and continue to increase. Although less abundant, these gases trap solar radiation much more effectively than carbon dioxide, and they are by now contributing about 45% to the greenhouse effect. Since pre-industrial times, green-house gases taken together have increased by an amount that is already equivalent, in its effect on trapping radiation, to a 50% increase in carbon dioxide. 'Effective doubling' of carbon dioxide – an increase of gases to a level that has an effect equivalent to a doubling of carbon dioxide – is expected by about 2020 on the business-as-usual scenario. By 2100, on this scenario, the total effect will be equivalent to 1200 ppmv of carbon dioxide, more than an effective quadrupling.

On the other emissions scenarios, the concentrations of

gases increase less rapidly. On Scenario D, they reach the effective doubling level only by 2100.

The IPCC does not offer projections beyond 2100. But its projections to 2100 show the effective carbon dioxide concentration still increasing at that date in all scenarios but D, and still increasing rapidly in the business-as-usual scenario. (Scenario D was developed particularly to show that levels of greenhouse gases could be brought under control by means of very tight restrictions.) Except for methane, the greenhouse gases persist in the atmosphere for a long time: 50 to 200 years on average for carbon dioxide; 150 years for nitrous oxide; 65 to 130 years for chlorofluoro-carbons.[5] The effects of these gases will therefore continue far beyond the next century.

All the IPCC projections ignore a wide range of feedback effects on the concentrations of greenhouse gases, mostly resulting from global warming itself. These feedback processes are not well understood, and cannot yet be quantified. There are very many potential feedback processes, nearly all of which are positive. Nearly all, that is to say, will tend to increase the concentrations of greenhouse gases.[6] One of several mentioned in the *Policymakers' Summary* is this. Cold water sinking from the surface of the oceans, particularly in the North Atlantic, carries carbon dioxide with it to the depths. This process removes a large amount of carbon dioxide from the atmosphere. As the earth warms and polar ice melts, surface water in the oceans will become less salty. This will reduce its density and is likely to slow the formation of deep water. It is likely, therefore, to reduce this sink for carbon dioxide.

Taking all the feedback processes together, the IPCC concludes that:

> Although many of these feedback processes are poorly understood, it seems likely that, overall, they will act to increase, rather than decrease, greenhouse gas concentrations in a warmer world.[7]

1.2   Climate and sea level

Greenhouse gases influence climate in extremely complex ways. The best instruments available for predicting their effects are *general circulation models* (GCMs). GCMs model the behaviour of the atmosphere according to the laws of physics, taking account of as many influences as possible. Each model works with a grid of points spread across the entire globe. The equations of the model are solved at these points only. Because the models make huge demands on computing power, the grid cannot be very fine. 'High-resolution' models use a horizontal spacing as wide as a few hundred kilometres.[8] But some important atmospheric phenomena take place on smaller scales than this. One is the formation of clouds. Cloud cover is a vital ingredient in predictions of global warming; clouds may either cool the world by reflecting radiation from the sun back to space, or warm it by reflecting radiation from the earth's surface back to the surface again. A process known as 'parameterization' has to be used in GCMs to take account of small-scale phenomena such as clouds. This process is open to controversy, and leads to large uncertainties in the models' results.[9]

Nevertheless, GCMs perform quite well in the tests to which they can be subjected. For instance, they successfully simulate the differences between winter and summer climates – differences that are much larger than the changes expected from global warming. For this reason, the IPCC has 'substantial confidence that models can predict at least the broad-scale features of climate change',[10] and bases its predictions on them. But because of the uncertainties, it offers a range of temperature estimates from low to high, with a 'best estimate' in the middle.

Temperature

Each GCM predicts a value for a quantity known as *climate sensitivity*. Climate sensitivity is defined as the increase in the equilibrium temperature of surface air, averaged across the globe, that will result from the doubling of carbon dioxide. The models give values for climate sensitivity between 1.9° and 5.2°, clustered around 4°. Out of this range, the IPCC picks 2.5° as its best estimate.

It offers two reasons for choosing a value so low in the range. First: 'Recent studies using a more detailed but not necessarily more accurate representation of cloud processes give results in the lower half of this range.'[11] Second: the observed warming of the world in the past century (about 0.5°) also suggests a value in the low half of the range.[12] Most models predict a greater warming than has actually occurred.

Climate sensitivity measures the *equilibrium* change in temperature. Actual temperature will lag behind the equilibrium, because the surface layers of the oceans have a large thermal capacity and are slow to warm up. Having taken a value of climate sensitivity from the GCMs, the IPCC used a much simpler model of atmosphere and ocean circulation to predict the path of temperature changes through time.

On the business-as-usual scenario, the IPCC's best estimate of global temperature for 2050 is 2.7° above its level in 1765. For 2100 it is 4.2° above. Its low estimate gives figures of 1.9° and 2.8° respectively; its high estimate 4.0° and 6.2°. Scenario D gives a rise of 2.0° by 2100, on the best estimate.

For comparison, global temperature is now about 5° above its level at the peak of the last ice age.[13] So if business continues as usual, increases in temperature comparable to the earth's emergence from an ice age are expected in a century or so. The rate of increase will be much faster than the emergence from an ice age.[14]

The IPCC makes no predictions of temperatures after 2100. But its projections for all its scenarios, even scenario D, show temperatures still rising in 2100. This is inevitable because, even if there were no further increases in green-houses gases after 2100, equilibrium temperatures would still be above actual temperatures at that date. So actual temperatures would continue to rise towards their equilibrium.

One of the IPCC's remark about the more distant future is significant. Climate is subject to natural variability. But the IPCC says that 'on a century time-scale this [variability] would be less than changes expected from greenhouse gas increases'.[15] This means, for one thing, that the IPCC is not expecting the onset of a new ice age. At one time, the prospect of global warming was welcomed, because it might repel an ice age that would otherwise have descended on us. But that no longer seems likely.[16]

Greenhouse gas feedbacks

I must emphasize that none of the above predictions takes account of the 'greenhouse gas feedbacks' mentioned in Section 1.1. Since these feedbacks are likely to be positive, the IPCC concludes that 'climate change is likely to be greater than the estimates we have given.'[17] That is to say, what the IPCC calls its 'best estimate' is not actually what it thinks is most likely to happen. It believes that the temperature is most likely to change by a greater amount, but it is not able to quantify this amount.

The *Policymakers' Summary* makes a particular mention of one feedback process. After 2100, it says, 'there could be significant changes in the ocean circulation, including a decrease in North Atlantic deep water formation'.[18] This is a reference to one of the major positive feedback process on the concentrations of carbon dioxide. I described it on page 4 above. Evidently the IPCC is concerned that this process might cause greater global warming in the further future.

Its concern does not stop at that. The circulation of the oceans may be a major controlling influence on climate. It may be, for instance, that ice ages come and go as a result of flips in the pattern of circulation between one stable state and another (caused initially by variations in the earth's orbit).[19] This may in turn affect the rate at which carbon dioxide is drawn from the atmosphere into the oceans. If global warming were to cause a flip to a different pattern, the consequences could be dire. The IPCC issues this warning:

> We must recognize that our imperfect understanding of climate processes (and corresponding inability to model them) could make us vulnerable to surprises; just as the human-made ozone hole over Antarctica was entirely unpredicted. In particular, the ocean circulation, changes in which are thought to have led to periods of comparatively rapid climate change at the end of the last ice age, is not well observed, understood or modelled.[20]

Other climate variables

The IPCC commits itself to few predictions about other global climate variables besides temperature. It does expect an increase in global average precipitation of a few percent by 2030. It finds no clear evidence that the climate will become more variable, nor that there will be an increase in the number of tropical storms, or a decrease in the number of temperate storms.[21]

It does offer, with less confidence than for global changes, a few predictions about regional changes. For instance, it predicts that, in the summer, central North America will have less precipitation and drier soils (15% to 20% drier in 2030) than at present.

Sea level

Global warming will cause average sea levels to rise during the next century for two main reasons. First, the warming of water in the top layer of the oceans causes it to expand. Second, ice on land in temperate regions will melt. The IPCC's best estimate under the business-as-usual scenario has sea levels rising 30 cm by 2050 and 65 cm by 2100. For the low estimate, the figures are 15 cm and 30 cm; for the high estimate, 50 cm and 110 cm. Under scenario D the best estimate rise is 32 cm by 2100.

A source of uncertainty, not included in these estimates, is Antarctic ice. If the West Antarctic Ice Sheet were to melt, that would increase global sea levels by about five meters. However, the IPCC does not expect this to occur in the next century.

Nevertheless, much larger rises in sea level are expected after 2100, even if there is no further increase in greenhouse gases. Global warming will raise equilibrium levels substantially, and it is only the delayed responses of the oceans and ice masses that will keep increases to comparatively modest levels during the next century.[22]

1.3 Societies and economies

What effects will global warming have on human life? The second of the IPCC's three working groups was asked to assess the environmental and socio-economic impacts of climate change. It has to be said that this group's conclusions are unclear and unspecific.[23] I have found in the literature a number of predictions about individual effects,[24] but no attempt besides the IPCC's at an overall assessment.[25] This paucity of conclusions is not surprising, since an accurate overall assessment cannot possibly be made. The effects of global warming on human life are necessarily very unpredictable. Global warming is disconcerting in one respect. It seems inevitable that its consequences will be large, just because of the unprecedented size and speed of the temperature changes. Yet it is very hard to know just what these large consequences will be.

Uncertainty

One reason is the very large uncertainties inherent in the predictions of future climate itself. I mentioned some of these in Sections 1.1 and 1.2. The main source of uncertainties is the greenhouse gas feedbacks, which are poorly understood and may be very large. But even ignoring these feedbacks, the IPCC's high estimate for temperature is more than twice its low estimate even by 2050, on the business-as-usual scenario.

But there are even more profound uncertainties involved in translating predictions about climate into effects on society. The difficulty is that, over the periods we are concerned with, society is bound to be radically transformed in ways that are utterly unpredictable to us now. In his paper 'Climatic change', Thomas Schelling makes this point graphically by asking us to imagine ourselves back to 1900. He asks us to imagine trying to make predictions at that time about the effects of climate on society at the end of the twentieth century. In 1900, he reminds us:

> Electronics was not dreamed of. . . . Transatlantic travel by zeppelin was a generation in the future. . . Russia was czarist. . . U. S. life expectancy at birth was 47. . . Only a third of the U. S. population lived in places with more than 5,000 inhabitants. . .

And so on. In 1900, looking forward, we would not have been able to imagine what society would be like in 1992, and we would therefore have been totally at sea in trying to predict the influence of climate on society. We are in a similar position now when we look forward to 2100. Let us call uncertainties of this sort 'historical uncertainties'.

Historical uncertainties make it very difficult to predict the effects of climate change. But it is also important to realize that they are themselves influenced by climate. Climate helps to determine history. Consider, for instance, the cooling that occurred at the end of medieval times, ushering in the Little Ice Age. This was a global cooling of only about 1°,[26] much less than the present predicted human-induced global warming. One of its effects was to cause the Vikings to retreat from Greenland. But for this event, North America might well have been colonized by Danes from the north, rather than by Spaniards from the south. And who is to know what the effect of that might have been?

It may be that nowadays technology has made mankind more independent of climate. But we do not yet have technological fixes for, say, desertification in Africa. And in any case, technology has little to do with some of the influences climate has on history. Large migrations in the United States, for instance, are caused simply by people's yearning for a sunnier climate. Despite the progress of technology, then, climate will still have a major effect on history.

A feature of historical uncertainty is that it may never be resolved. Not only are we now unable to predict what the effects of climate will be, but historians in the future will never know what many of its effects have been. No doubt the Little Ice Age had a vast influence on human history. But we do not know what would have happened if it had not occurred. So we do not know what its influence was, and we have no idea whether it was for good or harm.

Historical uncertainty leads Schelling to conclude:

> If we had perfect climate forecasts for all the inhabited regions of the world for the century that begins, say, in the year 2025, there would undoubtedly be important parts of the world, and segments of populations in all parts of the world, where it would be difficult to put an algebraic sign on the apparent welfare impact, let alone assess the magnitude. For the world as a whole we might not be confident of the direction of change in some aggregate measure of welfare. Undoubtedly there will be places where some predicted change in climate could have no foreseeable benefit and where some potential damages could be foreseen with clarity. But unless we impute to ourselves foresight much superior to what we might willingly claim for ourselves were we doing our work in 1900, it is likely that most of the identifiable changes in welfare due to climate change would be, for most parts of the world, swamped by other uncertainties.[27]

Nevertheless, I believe there are some general predictions

that we can make fairly safely, and that help set the programme for this study. The first is simply that the effects of human-induced global warming are very uncertain. As I shall explain, this by itself has important consequences for the work that needs to be done. But, more than that, we can say that the effects will certainly be long lived, almost certainly large, probably bad, and possibly disastrous.

Persistence

The evidence that the effects will be long lived is straightforward. Many greenhouse gases, including carbon dioxide, persist in the atmosphere for a century or more. This alone means that even the gases already artificially added to the atmosphere will influence temperatures for a long time. Furthermore, the slow warming of the ocean delays the full effect. Consequently, even if artificial emissions are stabilized by 2100, temperatures will continue to increase beyond that time. This is what the IPCC's scenario D demonstrates. Beyond this, the slow melting of the Antarctic Ice Sheet will prolong the effects for centuries, indeed for thousands of years.[28] It is also likely that positive feedbacks will be set off by global warming, which will prolong the effects. The effects on climate and sea level, indeed, may fairly be said to be irreversible on the time scale of human history. And this is just the effects on the natural world. The effects on human life, like the effects of the Little Ice Age, will persist through the rest of human history.

Size

The evidence that the effects will be large is a comparison with previous climatic changes. Between the peak of the last ice age and the present the earth has warmed by about 5°. The expected warming in the next century has a comparable magnitude, and it will happen much more quickly. The present interglacial period has so far lasted about ten thousand years, and during that time human civilization has come into being. Temperature fluctuations during this period have been only about 1°.[29] Temperatures during the previous interglacial period, about 130,000 years ago, were similar to today's.[30] This means that, within a few decades, the global average temperature will be higher than it has been at any time since *Homo sapiens* first evolved. It is inconceivable that changes on this scale could fail to have a profound influence on human life. The Little Ice Age resulted from a fluctuation of only 1°. The consequences will fail to be large only if the IPCC's projections for climate change, which are backed by a very wide consensus amongst scientists, turn out to be exaggerated.[31]

Direction

In the passage I quoted above, Schelling concludes from the historical uncertainties that we cannot even be confident about whether global warming will increase or decrease aggregate welfare. I, however, am willing to predict a decrease, at least for a century or so. In one respect our situation is now different from Schelling's. Schelling, writing in the early 1980s, was not expecting significant global warming to occur for many decades. Our present evidence, described in Sections 1.1 and 1.2 shortens the time scale. Very significant warming can now be expected within the lives of many people now living. This slightly reduces the uncertainties that impressed Schelling; we can expect important effects before historical uncertainties become imponderable. In any case, Schelling himself says: 'There is no reason for believing that the development is to be welcomed, and there are many reasons for the contrary.'[32] I agree. I offer the following three considerations in support.

   First, the strains on natural ecologies are likely to be very great. As the ice retreated at the end of the latest ice age, forests migrated northwards at perhaps 1 km per year. This appears to be about the maximum they are capable of in uncultivated country, and they will certainly not be able to manage the much faster movements required by the present global warming.[33] Furthermore, many ecosystems have become isolated by human activities, so they will only be able to migrate much more slowly, if at all. The natural world is therefore likely to be very much impoverished. And this will impoverish humanity. One might hope that the progress of technology has made agriculture more independent of the natural world: agriculture can migrate faster than natural ecosystems, and new crops can be matched to new conditions. But it seems overoptimistic to believe that agriculture can be restructured on this scale throughout the world without major costs. And in any case, we all need the natural world around us to make our lives rich and worthwhile. Life will not be so good in a more barren world.

Second, a number of consequences of global warming are more easily predictable than others, and these ones seem unambiguously bad. Chief amongst them are the effects of rising sea levels.[34] Without increased sea defences, low-lying areas will become more susceptible to flooding. The danger will be amplified if storms become more frequent or more severe, and there are some grounds for thinking this might happen,[35] though these grounds are not endorsed by the IPCC.[36] Regions threatened by flooding include densely populated areas. Eight to ten million people live within one metre of high tide in each of the unprotected river deltas of Bangladesh, Egypt and Vietnam.[37] A flood in Bangladesh, caused by a tropical storm in 1970, killed 300,000 people.[38] Rising sea levels, therefore, must be expected to kill very large numbers of people. This is an enormous and easily predictable harm that will be caused by global warming. Moreover, sea levels will continue to rise for centuries.[39] This must cause large migrations of population, and it is hard to see where the people can move to. There seems to be inevitable harm in this too: the forced migration of many million people is inevitably a disaster. Another class of bad effects is also quite easily predictable. As the world warms, more people will become subject to tropical diseases.[40] This, too, will shorten many people's lives.

Third, there are a number of specific predictions about

regional climate changes that suggest imminent harm to agriculture. It is the effects on water resources that seem most significant.[41] We need to be particularly cautious about regional effects, because the regional predictions of the climate models are still unreliable. Nevertheless, the projected summer drying of central North America, where much of the world's food supply is grown, can be identified as a serious threat.[42] The IPCC's working group on impacts suggested that there might also be some *benefits* to agriculture in some regions. But it expressed concern that already vulnerable areas might suffer particularly. The Sahel is one example.[43]

Unless there is a great ecological catastrophe, however, most of the harms that one can foresee from global warming could be classified as adjustment costs. I can see no reason why, in equilibrium, a warmer world should not be able to sustain just as good human life as a cooler one. The problem is that, over thousands of years, human beings and nature have become adjusted to a cooler world. Now, there is in fact no equilibrium in sight, so this thought gives no solid grounds for long-term optimism. But, combined with Schelling's historical uncertainty, which becomes very pressing beyond a century or so, it does mean we can be less sure that global warming will be harmful in the very long run. It also means that the harm done in the shorter run depends more on the *rate* of global warming than on the total *amount* of it. The IPCC concludes:

> What we can say with confidence is that the severity of the impacts will depend to a very large degree on the rate of climate change.[44]

Catastrophe

On page 8, I quoted the IPCC's cautiously-worded warning that poorly-understood feedback processes make us vulnerable to surprises. Some of these processes could be very powerful. The arctic methane hydrates, for instance, form a vast reservoir of methane,[45] and if a significant fraction of them was to be released, the earth might become uninhabitable. Human-induced global warming, then, could possibly start a chain of events that could lead to the extinction of civilization or even of humanity. This is a remote possibility, but it exists.

1.4 Implications

What conclusions can be drawn from my survey of predictions? We are looking for conclusions that will help point the direction for the rest of this study, and for future research.

Governments must act

My first conclusion is that national governments must take some action. Whatever was the basis of decisions that have been taken in the past, whatever the criteria that have been used for private and public investment, and whatever the grounds for existing national tax structures, none of them have so far taken account of global warming. The scientific predictions are new information that must alter the marginal balance between conflicting considerations in energy policy and elsewhere. So certain actions that previously appeared wrong will now turn out to be right. The question is not whether there should be action, but how much and what sort.

It must be *governments* that take action, because the greenhouse gases are public bads *par excellence*. No individual firm will benefit noticeably from reducing its own emission of carbon dioxide. Even actions such as raising sea defences, which are intended to mitigate the consequences of global warming rather than limit global warming itself, are very large-scale public goods. Of course, the right action for a government to take is not necessarily to spend large amounts of money on its own account. It might be better to alter the structure of the incentives that face individuals, by changes in the tax system or in other ways. But governments must do something.

It has sometimes been argued[46] that the uncertainty of the scientists' predictions is a reason for not acting at present, and that we should wait until some further research has been concluded. This argument is poor economics. The economic theory here is well established. It is nicely presented by Robert Lind in 'A primer on the major issues relating to the discount rate'. Since the application to the context of global warming is straightforward, I shall take the next four paragraphs to explain it here.

I believe that the effects of the greenhouse gases are probably bad. That is to say, although the harm done may be positive or negative, the expected value of harm is positive. But let us suppose, for the sake of argument, that the expected harm is zero. Consider some proposed project that would reduce emissions of carbon dioxide, at a cost in resources. Because we are supposing that the expected harm of carbon dioxide is zero, the expected benefit of this project is zero. This by itself may suggest that resources should not be used on it: why give resources to a project with no expected benefit? Furthermore, in a sense it is a risky project: its benefits are unknown, and they may even be negative. If carbon dioxide turns out to be harmful, then the project is beneficial, but if carbon dioxide is beneficial, the project is harmful. We generally assume that risk is to be avoided. If it is, that seems to suggest that the riskiness of this project is a further reason not to give resources to it.

However, what really matters is not the riskiness of the project considered in isolation, but the effect it has on the overall risk we are exposed to. Our emissions of carbon dioxide expose us to risk, because (we are supposing) they may be beneficial or they may be harmful. A project that reduces emissions reduces this risk; it reduces the variance in what may happen. It makes us less unsure about the future; it acts like insurance. If risk is a bad thing, that gives this project value as insurance, despite first appearances. This value must be compared with its cost in resources. If the cost is not too great, the project will be worthwhile.

Uncertainty about the effects of greenhouse gases is therefore a reason for devoting resources to reducing the concentrations of these gases, not a reason against it as is sometimes argued. The same point applies to projects aimed at mitigating the effects of global warming, such as raising sea defences. If there turns out not be any global warming, these projects will have been wasted. But they, too, act as insurance by reducing our overall risk. Any insurance premium is, in a sense, wasted if the insured risk does not come to pass. But insurance is nevertheless worthwhile.

All this is so even on the assumption that the expectation of harm from greenhouse gases is zero. Actually, though, I think the expected harm is positive. This is a further reason for acting to forestall it.

Uncertainty is central

Governments must act, and that means they must decide what actions to take. They must act now, despite the uncertainty of the consequences, and indeed in a sense partly because of the uncertainty. Their decision making therefore cannot follow the pattern: first find out the facts and then act on them. Uncertainty is an inherent part of the problem. Intuitively, it strongly conditions the nature of the problem. In particular, the small chance that global warming might lead to disaster seems, intuitively, to be one thing we ought never to lose sight of. More than anything else, most people are worried by the thought that we are interfering with the natural working of the entire globe, without properly understanding what we are doing. Our lack of knowledge has to be granted a recognized place within the decision-making process.

Conventional decision theory (expected utility theory) provides the standard account of how this is to be done. Roughly, it says: do whatever maximizes the expectation of utility.[47] This formula leaves a lot open. It leaves open, for one thing, the question of how probabilities, which determine the expectation of utility, should be arrived at. Is there a satisfactory way, for instance, of deriving them from the conflicting opinions of the scientific community?[48] This study is not specifically concerned with uncertainty, so these are not my questions. In principle I favour conventional decision theory. Nevertheless, when it comes to global warming, I do not think the decision-making process can be simply a matter of calculating expected utilities and then going ahead. The problem is too big for that, and the uncertainties – particularly the historical uncertainties – too extreme. In practice, the problem is in the domain of politics. I do not believe, for instance, that anyone could at present produce a useful cost-benefit analysis of, say, converting cars to run on natural gas rather than petrol, taking proper account of the expected effects of greenhouse gases. Instead, the decisions are going to be arrived at by political debate, and not by the calculations of cost-benefit analysts.[49]

The aims of this study

All this has implications for the aims of this study. Since governments must act, research on intergenerational relations must be aimed ultimately at providing guidance on how to act. Nevertheless, I believe it would be wrong to adopt the narrow aim of developing some formula for cost-benefit analysis, which governments could simply apply. I shall not confine myself to deriving a discount rate from current economic theory. The uncertainties of the problem are enough to make that exercise pointless. Cost-benefit analysis, when faced with uncertainties as big as these, would simply be self-deception. And in any case, it could not be a successful exercise, because the issue of our responsibility to future generations is too poorly understood, and too little accommodated in the current economic theory.

I shall set myself a broader, more general aim: roughly, to investigate how benefits and harms coming at one time are to be weighed against benefits and harms at another. I shall not be so ambitious as to try and answer this question, but I shall try to set out a framework for approaching it. My work will be much more fundamental than simply applying current economic theory. I shall be looking at the foundations of the theory. I hope the framework I set up may help to develop the theory and guide people in thinking about the issues, and perhaps give some organization to the political debate. Eventually, the framework might be tightened up enough, and made quantitatively precise enough, to be used in cost-benefit analysis. But not yet. For the moment, though action is essential, it cannot be handed over to technically expert decision makers.

Cost-benefit analysis is, in any case, only one of the applications where benefits and harms at different times need to be weighed against each other. When we face up to global warming, it may well not be the most important application. Cost-benefit analysis is aimed at evaluating government investment projects. Certainly, there is a role for government investment in dealing with global warming: sea defences are one example. But much of the problem may be better handled in other ways.

For instance, the properties of greenhouse gases makes them ideal targets for Pigovian taxes. Because carbon dioxide is so long lived compared with the mixing rate of the atmosphere, a unit of carbon dioxide released anywhere in the world does exactly the same damage as a unit released anywhere else. Consequently, if only we could work out the marginal damage done by a unit of carbon dioxide, one single tax on emissions set at this level, throughout the world (but perhaps with transfers from rich to poor countries to offset the distributional effects), would be a theoretically ideal solution. With this tax set, government projects to reduce emissions would be redundant. So we need, not a series of cost-benefit analyses, but this one single figure for marginal damage. In this respect, greenhouse gases differ from other pollutants, and in this one respect the problem they pose is simpler. The damage done by sulphur, for example, depends on where it is released. So a uniform sulphur tax would be inappropriate.

One single figure for the marginal damage of carbon dioxide is, of course, appallingly difficult to arrive at. The marginal damage done by the gas will change over time, and, being a marginal quantity, it will change according to the amount of carbon dioxide already in the atmosphere. Another difficulty is the one this study is concerned with. The damage will come at various dates far into the future. So to fix on an overall figure requires us to assign relative weights to damage at different dates. This, then, is another application for the question of how benefits and harms at one time are to be weighed against benefits and harms at another.

Furthermore, we need marginal damage figures for all the other greenhouse gases as well. Different gases have different lifetimes. So comparing together the harms done by different gases is, once more, a matter of assigning weights to damage done at different future times. A quantity known as the *global warming potential* of a gas may be seen as a crude attempt at cutting through this complication. A gas's global warming potential is the total warming effect a unit of the gas produces during its lifetime. Methane, for instance, is a much more powerful greenhouse gas than carbon dioxide, but it has a shorter life. On balance, it has about nine times the global warming potential.[50] But global warming potential is a very crude index of the relative harmfulness of the gases, because it assigns the same weight to warming at one date as it does to warming at another. This cannot be correct. It cannot be correct even if – I shall be discussing this – harm done at one time should have the same weight as harm done at another. The connection between warming and harm is a very complex one, and warming at one time will certainly not be equally as harmful as warming at another time. One obvious reason is that the population of the world is changing over time, so warming at different times, whatever harm it does, imposes its harms on different numbers of people. Once again, we cannot avoid careful work on the weighing of harms at different dates.

Our general problem, then, is weighing harms and benefits at different times. When we confront global warming, this problem raises its head in various different concrete applications. One is cost-benefit analysis. Another is estimating the marginal damage of greenhouse gases. And there are others too.

Distribution within a lifetime and between generations

My survey of the state of science has three other implications that need to be mentioned. The first is that, with such a long time scale, the problem of weighing harms and benefits at different times may fairly be called 'intergenerational'. It involves our responsibility to future generations. Since the effects of global warming will be very long lived indeed, some of these generations are very far in the future.

On the other hand, it also needs to be said that a part of the problem already lies within a generation. Substantial global warming is expected in just a few decades. Harms to people now alive are amongst the most predictable: many of them will be killed by floods. In contrast, the effects on the more distant future are shrouded in historical uncertainty. We cannot even be sure they will be bad. In this respect, the problem of global warming differs from some other long-term problems: nuclear waste, for instance. Whatever the course of history, we can be almost sure that, unless properly shielded, nuclear waste will go on doing harm for many thousands of years. We cannot be so sure about global warming.

In considering how to weigh the future against the present, then, we need to consider both weighing within a generation, and weighing between generations.

The international perspective

Greenhouse gases are the most perfect examples of public
bads: they are global bads. Only the biggest nations will be
much affected by their own emissions. If nations act separately to promote their own interests, then each will find it
in its own interest to ride free. Emissions will be reduced by
far less than they ought to be. Consequently, unless there is
to be international cooperation over the problem, it is
pointless to debate about what ought to be done. I do not
wish to participate in a pointless debate. I shall therefore
take as my aim in this study to consider what ought to be
done from a global point of view, not a national one.

Demographic effects

Global warming will, no doubt, make some people's lives
worse than they otherwise would have been, without
changing the length of those lives. Some of its harmful
effects will be of this sort. But it will also shorten people's
lives. The very many deaths that will be caused by flooding
and diseases are amongst its most predictable consequences.
Furthermore, it will force large movements of population,
and inevitably cause large changes in the size of populations. That is to say, it will affect the numbers of people who
come into existence. Judgements about the harms (and
benefits) of global warming will therefore have to take into
account changes of populations, and the lengths of people's
lives. Let us call these 'demographic effects'. We need to be
able to assign a value to demographic effects.

The most extreme possible demographic effect is that
global warming will prevent the existence of any people at
all after some date in the future. If positive feedbacks get
out of control, the earth may become uninhabitable by
human beings. No doubt, the chance that this will happen
is very small. On the other hand, if it does happen, the
consequence is so extreme that this small chance is still an
important consideration. Attention needs to be given to the

harm of extinction, and how to accommodate the possibility of extinction in our decision making. I regret to say, however, that I have not been able to consider this problem in this study.[51]

Notes

1. Houghton, Jenkins and Ephraums, *Climate Change*.
2. Intergovernmental Panel on Climate Change, *Scientific Assessment of Climate Change: The Policymakers' Summary*.
3. *Policymakers' Summary*, Foreword.
4. p. 26.
5. p. 7.
6. In 'The nature of the greenhouse threat', p. 30, Leggett lists the feedback processes that were examined by the IPCC. Of these at least fourteen are not accounted for in the IPCC projections we have mentioned, and only one of these is known to be negative.
7. *Policymakers' Summary*, p. 10.
8. Schneider, 'The science of climate-modelling', p. 48.
9. In 'The nature of the greenhouse threat', p. 31, Leggett reports an example. The UK Meteorological Office's model at one time predicted that effective doubling of carbon dioxide would warm the world by 5.5°. A change in the method of parameterizing clouds reduced this prediction to 1.9°.
10. *Policymakers' Summary*, p. 19.
11. *Policymakers' Summary*, p. 15. Notice the strained logic in this remark. If the IPCC does not believe that the recent studies are necessarily more accurate, why does it use them as grounds for fixing on a low value of climate sensitivity? I suspect the IPCC is being conservative, for the sake of achieving consensus.
12. *Policymakers' Summary*, p. 21.
13. Schneider, 'The changing climate', p. 43.
14. In 'The changing climate', p. 43, Schneider estimates ten to a hundred times faster.
15. *Policymakers' Summary*, p. 19.
16. In 1938, Callendar welcomed the greenhouse effect because it would hold 'the deadly glaciers' at bay ('The artificial production of carbon dioxide', quoted by Jamieson, 'Managing the future'). But present understanding is that the cycle of glaciation is controlled by features of the earth's orbit round the sun, and these features would by themselves only cause very small changes in global temperature during the next five thousand years or so. (See Wigley, Jones and Kelly, 'Warm world scenarios', pp. 276–7.)
17. *Policymakers' Summary*, p. 19.
18. p. 19.

19. Broeker and Denton, 'What drives glacial cycles?'
20. *Policymakers' Summary*, p. 19. A change in the circulation of the oceans is only one of several potentially very large feedback processes. Another is this. Under the arctic permafrost and around the edges of the Arctic Ocean are vast quantities of methane hydrates. These contain methane locked in a crystalline formation with ice. (See Schimel, 'Biogeochemical feedbacks', p. 74.) Methane is a potent greenhouse gas. If human-induced global warming was to melt significant quantities of the methane hydrates, it could lead to runaway global heating.
21. For a contrary view, see Stark, 'Designing for coastal structures in a greenhouse age', and the references given there.
22. In 'The expected sea-level rise from climatic warming in the Antarctic', Budd predicts a new equilibrium sea level 4.5 m above the present level, caused by a melting of Antarctic ice. But he does not expect this level to be reached for five thousand years. The contribution of Antarctic ice to sea level in five hundred years he expects to be one metre.
23. See IPCC, *Overview and Conclusions*, Chapter 3.
24. E.g. Haines, 'The implications for health'; Schneider, 'The changing climate'; Woodwell, 'The effects of global warming'.
25. Since this report was written, Nordhaus has published an estimate of the impact on U. S. national income of an effective doubling of carbon dioxide ('To slow or not to slow'). His conclusion is briefly discussed in note 31 below.
26. IPCC, *Policymakers' Summary*, p. 20.
27. p. 454.
28. Budd, 'The expected sea-level rise from climatic warming in the Antarctic'.
29. IPCC, *Policymakers' Summary*, p. 20.
30. Wigley, Jones and Kelly, 'Warm world scenarios', p. 274.
31. In 'To slow or not to slow', published since this report was written, Nordhaus suggests that effective carbon dioxide doubling will diminish U. S. national income by only 0.25%. The paper does not give details of his calculation, but I have one comment to make. Nordhaus's estimate appears to be based on the assumption that everything will be much as it is now, but a bit hotter. For instance, he appears to estimate the impact on agriculture by calculating the effect of higher temperatures and higher concentrations of carbon dioxide on the yield of existing crops. However, the impact on natural ecologies of such an unprecedented change in climate cannot fail to be very great. So Nordhaus is evidently assuming that human life is by now fairly independent of the natural world. As I shall be saying on page 14, I find this assumption too complacent. Nordhaus would presumably have produced a similar estimate for the effect on national income of a similar global cooling. But a similar cooling would take us most of the way to an ice age, and I doubt if anyone would expect that to diminish national income by a mere .25%. I think we must expect global

warming to have a profound effect on history, rather than a negligible effect on national income. In 'Global warming', also published after this report was written, Beckerman takes a similar view to Nordhaus's.

32. p. 480.

33. Huntley, 'Lessons from climates of the past'.

34. IPCC, *Overview and Conclusions*, pp. 9–11.

35. See Stark, 'Designing for coastal structures in a greenhouse age', p. 165.

36. *Scientific Assessment: Policymakers' Summary*, p. 17.

37. IPCC, *Policymakers' Summary: Working Group III*, p. 14.

38. Berz, 'Climatic change: impact on international reinsurance', p. 586.

39. See page 9 above.

40. *Overview and Conclusions*, p. 13; Haines, 'The implications for health'.

41. *Overview and Conclusions*, pp. 11–13.

42. Schneider, 'The changing climate', p. 45.

43. *Overview and Conclusions*, pp. 11–12.

44. *Overview and Conclusions*, p. 9.

45. Leggett, 'The nature of the greenhouse threat', p. 40.

46. For instance, by John Sununu, President Bush's Chief of Staff. See *Scientific American* (April 1991) pp. 16–17.

47. 'Utility' is a technical term, and utility is not necessarily the same as benefit. Expected utility theory does not say: do whatever maximizes the expectation of benefit. See my *Weighing Goods*, Chapter 6.

48. There is a survey of the literature on this problem in French, 'Group consensus using expert opinions'. An example of more recent work is Wagner, 'Consensus for belief functions'.

49. For a further discussion, see Jamieson, 'Managing the future'.

50. IPCC, *Policymakers' Summary*, pp. 11–12. The IPCC's use of global warming potential is actually more complicated than I have suggested. It defines global warming potential relative to a time horizon. A gas's global warming potential is its total warming effect over the first T years of its life, where T is the time horizon. The figure I have given of nine for methane relative to carbon dioxide is for a five-hundred year horizon.

51. There is one remark about it on p. 121. The problem has been considered in more depth by other authors. For instance: Parfit, *Reasons and Persons*, pp. 453–4; Sikora, 'Is it wrong to prevent the existence of future generations?'. For a discussion of how to accommodate the possibility of disaster into decision making, see Morton, *Disasters and Dilemmas*.

Chapter 2
Justice and Wellbeing

What should be done about global warming? I propose to approach this question in terms of benefits and harms, or more accurately goods and bads. Specifically, I shall consider how goods and bads coming at different times are to be weighed against each other; often I shall consider how goods coming to one generation are to be weighed against goods coming to another. Suppose some action is in prospect, such as building up the nuclear energy programme. It will bring some benefits and it will do some harms. If the benefits outweigh the harms, I shall take the action to be a good one. More accurately, I shall take it to be better than the alternative of not doing it; I shall be concerned only with the relative goodness of actions – with whether one is better or worse than another – and not with any sort of absolute goodness.

When there is a range of alternative actions, one of them will turn out to be the best, after weighing up goods and bads (or perhaps several might be equally best). It is natural to suppose that this is the one that ought to be done (or that one of the equal best ought to be done). This view is known as *teleology*, or sometimes *consequentialism*. Teleology is the view that the right action – the action that ought to be done – is necessarily the best of the alternatives available.

Teleology, then, judges actions exclusively by goodness. Suppose the question is whether we should limit emissions of carbon dioxide. According to teleology, the answer is a matter of whether it is better to limit emissions or not, and this will be a matter of weighing against each other the benefits and sacrifices of different generations.

I do not insist on the complete truth of teleology. There may perhaps be other considerations besides teleological ones. But I do believe that teleology holds a large part of the truth, enough to justify me in conducting this study in teleological terms. I believe that the weighing up of goods and bads goes, at least, a long way towards determining what ought to be done. This is so only because I interpret the notion of good very broadly, as I shall explain in Section 2.2.

But before that, I need to mention a number of views that conflict with teleology. Many nonteleological views are centred around the notions of justice and rights, and many of the arguments about our responsibilities to future generations are conducted in these terms. It is said that future generations have a right to a share of the earth's natural resources, that justice requires us not to harm them by polluting the atmosphere, and so on. According to views like this, weighing up goods and bads is not enough. We would not, for instance, be entitled to impose some harm on a future generation, just because we have calculated that the harm is outweighed by some greater benefit to ourselves.

I do not wish to argue against nonteleological views about justice and rights in general. Perhaps they work well in other contexts. The objection I shall raise here is that, although they have been applied to relations between generations, they do not cope with them well. That is the subject of Section 2.1.

Section 2.2 explains the idea of teleology a little further, and Section 2.3 sets out in more detail the framework of my proposed teleological analysis.

2.1 Justice between generations

The idea of justice is a broad one, and I am not so foolish as to deny that relations between generations involve justice in some sense. But some strands in modern thinking about justice are inimical to teleology. I believe, however, that these particular strands of thought have special difficulties when applied across generations, whatever their merits in general. In this section I shall describe four of these ideas, and explain why I think they do not cope well with intergenerational justice.

Contractualism

The recent spate of interest in nonteleological notions of justice stems from John Rawls's *A Theory of Justice*. Rawls's theory gives priority to the right over the good, as he puts it. It is intended as an alternative to utilitarianism, which, as a version of teleology, gives priority to the good over the right. Rawls does not set up a notion of good in advance, and then ask how to achieve it. Instead he asks how people in society, who may have quite different and conflicting conceptions of what is good, can come together to organize their social life. Lawlessness is bad for everyone; if each person tried to pursue good as she conceives it without constraint, life would be nasty, brutish and short. Everyone benefits from accepting rules that constrain each person's conduct. 'Justice' is the name we give to the rules that regulate the conduct of people in a properly organized society. The idea of justice, then, is derived from people's mutual advantage, and it is to everyone's advantage to accept rules of justice. As a device for investigating what the principles of justice are, Rawls imagines people coming together in an 'original position', where they agree upon what principles they will live by. The broad school of thought that has followed Rawls's work I shall call 'contractualist'. Because it is not interested in the pursuit of good, contractualism is opposed to teleology.

The contractualist idea of justice runs into a difficulty when it comes to justice between generations, one that has been convincingly exposed by Brian Barry.[1] Rawls takes over from David Hume the doctrine that the idea of justice can only arise in particular circumstances.[2] It arises when people are roughly equal in power but limited in their mutual generosity, and face conditions of moderate scarcity. Principles of justice can regulate their interaction in such circumstances. Rough equality is required by the contractualist idea because, if one group of people has complete power over another, the powerful group has no reason to accept any rule that restricts its conduct towards the other. So if one group of people has absolute dominion over another, no question of justice arises between them. Hume says:

> Were there a species of creature intermingled with men which, though rational, were possessed of such inferior strength, both of body and mind, that they were incapable of all resistance and could never, upon the highest provocation, make us feel the effects of their resentment, the necessary consequence, I think, is that we should be bound, by the laws of humanity, to give gentle usage to these creatures, but should not, properly speaking, lie under any restraint of justice with regard to them, nor could they possess any right or property exclusive of such arbitrary lords. Our intercourse with them could not be called society, which supposes a degree of equality, but absolute command on one side and servile obedience on the other. Whatever we covet, they must instantly resign. Our permission is the only tenure by which they hold their possessions, our compassion and kindness the only check by which they curb our lawless will; and as no inconvenience ever results from the exercise of a power so firmly established in nature, the restraints of justice and property, being totally useless, would never have place in so unequal a confederacy.[3]

In our society, people not yet born are in exactly the position of Hume's weakly creatures. Whatever we covet from the earth's resources, they must instantly resign, and our permission is the only tenure by which they hold their share of these resources. According to Hume's doctrine, therefore, no question of justice can arise between us and future generations. If justice is simply the principles accepted by people for their mutual advantage in regulating society, then there is no such thing as intergenerational justice.

Intergenerational justice can only be got off the ground by moving away from Hume's circumstances of justice. Rawls himself fixes specific principles of justice by imagining the deliberations of people placed behind a 'veil of ignorance'. These people are supposed to come to an agreement about the principles they will live by, without knowing their own eventual station in society. The veil of ignorance overcomes the problem raised by the unequal power of different generations.[4] People in the original position do not know which generation they will find themselves in. They will therefore agree upon principles that do not allow the first generation to grab all the resources of the world. Rawls believes they will fix on a 'just savings rate'.[5] But Barry points out that, when Rawls uses the veil of ignorance in this way, he is no longer treating justice as the rules that regulate society for people's mutual advantage.[6] One generation gains no advantage in making concessions to later generations. So Rawls has abandoned Hume's position.

A different development of Hume's idea is found in David Gauthier's *Morals by Agreement*. Gauthier uses no veil of ignorance, but supposes that the principles of justice are to be derived from bargaining between people who know very well their own situation. However, Gauthier constrains their bargaining by a 'proviso', drawn originally from John Locke. In the bargaining process, the proviso says one person is not allowed, or even allowed to threaten, to make another person worse off than she would have been in the first person's absence.[7] A generation, then, is not allowed to use up any of the earth's exhaustible resources, unless it compensates later generations in some way, perhaps by providing them with an improved technology. Otherwise it makes later generations worse off than they would have been in its absence. In Gauthier's treatment of justice between generations, this proviso does most of the work.[8]

Now, the principle that each generation should leave the next equally well placed is what other authors work hard to establish as a conclusion of their arguments about intergenerational justice.[9] But Gauthier assumes it at the start. Gauthier does his best to justify the proviso in general terms (not specifically between generations); he tries to show that rational people will accept it as a condition on their bargaining.[10] But however successful his arguments may be in general, they are much less convincing when the parties in the bargaining belong to different generations. The arguments are all concerned with the interaction of the parties, and how they benefit mutually from being nice to each other. But between generations there is not much interaction. It is true that each generation overlaps with some of its successors and some of its predecessors. Gauthier hopes that this overlap is enough to allow his arguments to extend to intergenerational justice.[11] But he makes no detailed attempt to justify this hope, and I doubt that he could really defend the proviso on these grounds. I do not think, then, that Gauthier's is a successful attempt to derive principles of intergenerational justice from the idea of mutual advantage.

All this suggests that, when it comes to justice between generations, the contractualist idea of justice as mutual advantage has to be tempered with some other outlook on justice. There may well be scope for dealing with the problem within the broad school of contractualism. Peter Laslett, for instance, favours the idea of a 'tricontract' involving three generations at a time: each generation has contractual duties towards the next generation, which are balanced by rights it holds over the previous generation.[12] Nevertheless, I think the gap in the idea of justice as mutual advantage should make contractualists more friendly towards teleology. It means they will have to import into their theory some considerations from elsewhere, and teleological considerations of good may well be what they need. Brian Barry argues that Rawls is anyway ambivalent between justice as mutual advantage and a quite different approach to the idea of justice, which Barry calls 'justice as impartiality'.[13] I am very comfortable with the idea of justice as impartiality, and it fits well into teleology. It is discussed in Section 3.4.

Rights

A correlate of the idea of justice is the idea of a right. What justice requires me to do is normally a duty owed to someone, and that person has a right that I do this thing. Justice, for instance, requires not to pollute with smoke the air you breathe; I owe it to you not to pollute your air, and you have a right that I do not do so. The idea of rights is inimical to teleology because it is often claimed that rights cannot be weighed against other goods. They are what Robert Nozick calls 'side constraints'.[14] Rights must be satisfied before we even think about maximizing people's wellbeing. It is often said about justice that it is absolute and not negotiable: 'Fiat justitia, ruat caelum'. This characteristic is hard to accommodate in teleology.

Now, suppose justice requires us not to pollute the atmosphere with greenhouse gases. It seems that this duty not to pollute must be owed to those people living in the future who will suffer if we do, and that those people have a right not to have their atmosphere polluted by us. However, it turns out to be difficult to sustain the idea that we owe such a duty to future people.

Compare what will happen if we take steps to control our pollution of the atmosphere with what will happen if we do not. The steps we shall have to take will make a significant difference to people's lives. In the rich countries, for instance, people will almost certainly have to travel about less. Consequently, young people will form different groups of friends, meet different people, and marry different people. They will have children at different times, and those will, of course, be different children. After a century or so, nearly all of the people then living will be different individuals from the people who will be living if we continue to pollute in our present profligate way.

Now, the appeal to rights is supposed to give us a reason to control pollution. The reason it offers is that future people have a right that we should do so. But which future people are those? They would have to be the people who would suffer from our pollution if we were not to control it. But if we control pollution, there will be no such people: there will be no people such that, if we were to pollute, they would suffer as a result. The people who would experience the consequences if we were to pollute would be quite different from the people who will exist if we control pollution. If we control pollution, then, there is no one to whom we owe it as a right that we should do so. So it cannot be a reason to control pollution that we owe it as a right to future people.

This conclusion is an aspect of what Derek Parfit calls 'the nonidentity problem'.[15] Thomas Schwartz believes it implies that we are under no obligation to provide any widespread, continuing benefit to our descendants.[16] But this is the wrong conclusion to draw. The right one, I believe, was drawn by Douglas MacLean:

> Why should the identity problem be seen as undermining a kind of moral responsibility, rather than simply one kind of moral argument? It leads directly to skepticism only for those who are convinced that intergenerational morality must be a matter of justice or rights.[17]

Our responsibilities to future generations are, I believe, more satisfactorily handled in terms of good than in terms of rights. The nonidentity problem is one of my reasons for thinking so. But I do not think it is a conclusive reason by itself. There is a strong intuitive appeal in the thought that we owe it to future generations to leave their atmosphere unpolluted, and that they have a right to unpolluted air. One way this thought might be rescued from the nonidentity problem is to recognize that the owners of rights are not necessarily individual people. It seems that nations have rights. Kuwait has a right not to have its territory seized, and this right seems separate from the rights of the individual citizens of Kuwait. It will survive even when the entire present population of Kuwait has died, and been replaced with a new population. Perhaps the rights of a generation might be conceived in a similar way.

Resourcism

Suppose we think that equality is an aspect of justice. Then the question arises: equality of what? Does justice require each person to have the same wellbeing as others? Or should there be equality of something other than wellbeing? *Resourcism* is the view that justice is concerned with equality in the resources people have at their disposal, and not equality in wellbeing. The thought behind it is that the responsibility of society is only to equip people with the means to make something of their lives. What they do with those means is up to them.[18] It seems, then, that resourcism will oppose teleology. It is resources that matter to resourcism, whereas teleology is concerned with the benefit people derive from resources.

I am by no means opposed to resourcism, and I believe that a resourcist view of equality can be reconciled with teleology.[19] But I shall not try to make the reconciliation here. Instead, I want simply to say that, when it comes to relations between generations, the distance between resourcism – egalitarianism about resources – and egalitarianism about wellbeing is in practice small. To explain why, I shall take as my example Brian Barry's 'Intergenerational justice in energy policy', which is an application of resourcism to intergenerational justice.

Barry argues that each generation should have access to the same productive potential as any other. One aspect of this claim is its 'rigid egalitarianism', which I shall discuss below. Another is its resourcism. Economists interested in economic growth generally concern themselves with the

development of people's consumption over time. But Barry
thinks we should concern ourselves with the development of
the economy's productive capacity rather than consumption.
If some generation chooses not to work so hard as others,
and consequently consumes less, Barry thinks that is no
concern of justice.[20] Justice cannot require each generation
to leave exactly the same resources for the next generation
as it received itself, because then exhaustible resources
would never be used at all. But according to Barry, justice
does require that if a generation depletes some resources, it
should compensate its successors by supplying them with
productive capacity in other ways. For instance, it could
leave them extra capital. Barry, then, believes that productive capacity should be equalized across generations.

He must be wrong in one respect, though. Imagine a
programme for the development of the economy that gives
each generation the same consumption. Nothing says that,
in this programme, each generation will *save* the same
amount. Early generations, for instance, possessing better
natural resources, may well be required by the programme
to save a greater proportion of their production, in order to
build up capital for the future when resources are depleted.
Later generations may simply be required to maintain the
capital they inherit, and so save less. Since later generations
consume the same as earlier ones, but save less, they must
evidently produce less. Assume for simplicity that each
generation works just as hard as the others. Then later
generations evidently possess a smaller productive capacity
than earlier ones. Does this mean they are hard done by?
Could they complain of injustice? Plainly not. Earlier
generations possess a greater productive capacity, but a part
of it is committed to providing saving for the sake of later
generations. Saving is the means by which the fruits of
production are redistributed across generations. Plainly, an
egalitarian, even a resourcist, should not insist on equality
of productive capacity itself, but on equality of the part of
productive capacity that is not committed to redistribution.
To ask for equality of productive capacity, as Barry does, is

like asking an egalitarian tax system to equalize people's before-tax incomes rather than their after-tax incomes.

Even a resourcist, therefore, should not be concerned with the resources or productive capacity available to a generation, but with productive capacity *less* that generation's saving. Now compare the concern expressed in typical economic models of development. Economists are typically concerned for wellbeing, and if anything was to be equalized in their models, it would be wellbeing. But wellbeing is taken to depend only on consumption, and consumption is production less saving. So economists are concerned with production less saving. How does this concern differ from the resourcist's concern for productive *capacity* less saving? Not much. If each generation works the same amount, productive capacity maps directly into production, so that in practice the two amounts will be just the same. And there is no reason to think generations will differ significantly in the amount of work they do. Certainly, the models do not allow for such a difference. So equalizing consumption in the models will be the same as equalizing productive capacity less saving.

The place where resourcism makes an important difference is in distribution between individuals. Some individuals no doubt use their resources better than others. Barry says, of two people, that 'if we discover that one of them gets more fun out of spending his income than does the other, this is no reason for transferring income from the one who derives more utility to the one who derives less'.[21] He may be right: between individuals we should perhaps pursue equality of resources rather than wellbeing. Since individuals differ in their personalities – some enjoy life more than others – this makes a significant difference. But when it comes to distribution between generations, we have no need to allow for differences in personality. So resourcism and egalitarianism about wellbeing will in practice come to much the same thing.

Justice and Wellbeing

Rigid egalitarianism

Many modern theories of justice are associated with very rigid views about equality. One example is John Rawls's 'difference principle'.[22] Society should be arranged, Rawls thinks, so that the worst-off group is as well off as possible in terms of access to goods. This is not to insist on complete equality, but Rawls does insist that losses to the worst-off group can never be outweighed by gains to others. Rawls sets himself against weighing one person's gains or losses against another's. But the teleological process of maximizing good – picking the best of the alternatives – does typically involve this sort of weighing. So this strand of thinking, too, seems inimical to my teleological programme.

Applied to relations between generations, a rigidity of this sort appears in Barry's 'Intergenerational justice in energy policy'. Barry insists that each generation should have the same productive capacity. He would not countenance even a small reduction suffered by one generation for the sake of a great gain for another. This is a rigid egalitarianism about resources. Rigid egalitarianism, this time about wellbeing, is also to be found in Robert Solow's 'The economics of resources or the resources of economics'. A related but less rigid view forms one element of the popular idea of 'sustainable development'. David Pearce, Edward Barbier and Anil Markandya, for instance, favour sustainable development, and they understand it as the view that economic development should be constrained by the condition that natural resources must not decrease over time.[23] These authors have nothing against *increasing* natural resources, to allow future generations to enjoy more resources than we do. So long as natural resources never decrease, they are happy to allow one generation's wellbeing to be weighed against another's. But they do insist on the rigid, resourcist constraint that later generations should be bequeathed as least as many natural resources as earlier ones.

Rigid egalitarianism between generations seems to me

implausible. Interestingly enough, even Rawls evidently found it so, and could not bring himself to apply his difference principle between generations.[24] Rawls was writing at a time when economic growth seemed endless. The worst-off generation in these circumstances is the first. This generation could improve its wellbeing by investing less for the future. It would be as well off as possible, without making any other generation worse off than itself, if it invested so little that growth was abolished entirely. Then all generations would end up as badly off as the first. Rawls could not bring himself to propose, in the name of justice, that this should happen – that there should be no economic growth. Rawls was willing to accept that the first generation should be slightly worse off than it might be, for the sake of making later generations better off. So, in the context of intergenerational justice, even Rawls was willing to weigh a sacrifice to one generation against a gain to another.

Things are different now, and growth does not seem endless at all. Rigid egalitarianism may not now seem so obviously wrong. Now its implication may be that the current generation should be abstemious, whereas to Rawls egalitarianism suggested that the current generation should be profligate and save nothing for its successors. But the basis of the arguments remains the same. If weighing was tolerable when we expected endless growth, it is tolerable now.

Barry's argument is that each generation has the same claim on resources, and therefore each generation should have an equal share. 'Equal shares is the only solution compatible with justice.'[25] I agree that justice or (I would rather say) fairness, requires equal shares. But I do not think that the demands of fairness are absolute. It may be right sometimes to allow some unfairness, for the sake of greater overall benefit; I have argued this case elsewhere.[26] I shall be explaining on page 43 that I take fairness to be a sort of good. Teleological maximizing, as I conceive it, will give value to fairness, without taking up an extreme rigidity.

Barry has taken a very large leap from one extreme position to another. Earlier in this section we were facing the question of why one generation should sacrifice *anything* for later generations. Why should it not simply take all the resources it wants? Barry pointed out that Hume's doctrine about the circumstances of justice would allow it to do so. That is one extreme position. Suppose, abandoning that extreme, we grant that the present generation should not simply take what it wants. Then the next question is whether future generations should count *as much* as the present generation in our decisions, or whether it is permissible to discount their good compared with ours. In Chapter 3, I shall consider arguments on each side of this question. Next, going further, suppose we decide that all generations should count equally, and we should not discount the wellbeing of future generations. This is still not a view that is seriously egalitarian between generations. It values only the total wellbeing of all generations, and is not interested in how that wellbeing is distributed amongst the generations. If we wished, we could go still further and introduce egalitarian elements in various ways. One is the way I favour, of treating fairness as a good.[27] Another way, common in economics, is this. Instead of maximizing the total wellbeing of all generations, we apply a strictly concave transformation to the wellbeing of each generation, and maximize the total of the transforms.[28] This gives some weight to equality – for a given total of wellbeing, it is better to have it equally distributed amongst the generations rather than unequally – but it does not insist rigidly on equality. It allows one generation's wellbeing to be weighed against another's, and equality to be weighed against overall benefit. So there are various moderate egalitarian possibilities. But finally, passing over all of them, we come to the view that insists on rigid equality between generations. This is Barry's view, and it is the opposite extreme position.

I have seen no more moderate type of egalitarianism between generations explored anywhere.[29] This is a yawning gap in the literature. But I see no attraction in Barry's extreme position. I favour a position that does not give absolute precedence to equality, but allows it to be balanced against other values.

2.2 *A broad notion of wellbeing*

I intend to investigate relations between generations in teleological terms. So I should say a little more about what this means. It is difficult to draw a sharp line between teleology and other ethical theories, and I shall not try to draw the line precisely here.[30] I shall describe it only roughly.

Teleology is the theory that one should act so as to maximize good. The notion of good can include many ethical values. For instance, it can include the value of equality, either within a generation or between generations. If equality is a good thing, then that good can be put together with other goods, such as the total of people's wellbeing, to determine overall good. So a teleologist can aim to maximize overall good, taking equality into account.

If any ethical theory says we should maximize something, it is fair to say that that something is what the theory considers good. So any maximizing theory is teleological; it says we should maximize good, as the theory conceives good. Teleology can be identified by its *structure*, therefore: it has a maximizing structure. So long as a theory sets up an objective that should be maximized, it is a teleological theory. That is how I identify teleology.

What does this definition exclude from teleology? Probably not much. It may be that, with some work, almost any ethical theory could be cast in a maximizing form. Certainly, some theories are inimical to teleology on first appearances, at least. I have mentioned some in Section 2.1. One typical characteristic of teleology is that it permits the *weighing* of different goods against each other. It permits equality to be balanced against total wellbeing, for instance. For this reason, I described a side-constraint theory of rights, and also rigid egalitarianism, as inimical to teleology. These theories do not permit weighing. However, an objective can be designed that does not permit weighing either: a maximin objective is like this. And I call any theory teleological if it aims to maximize an objective. So it is possible that even these non-weighing theories could be fitted into teleology. But since, as I explained, I find these theories unconvincing when applied to relations between generations, I shall not trouble to do so.

I identify teleology by its structure, then, rather than by a particular notion of good. This implies, for one thing, that teleology is not what Amartya Sen calls 'welfarism'.[31] When there is a choice between actions, welfarism is the theory that, in determining which is right, the only thing that matters is the welfare that each action will bring to people. Sen understands 'welfare' very narrowly. For instance, treating a person unjustly he would not count as damaging her welfare, unless she feels bad about it. The bad feeling is a negative part of her welfare, but not the injustice itself. In his paper 'Approaches to the choice of discount rates', Sen argues that welfarism is inadequate for understanding relations between generations. He may be right, but it does not follow that teleology is inadequate.

My notion of good is much wider than Sen's notion of welfare. Sen himself, indeed, has helped to develop a broad notion of good. For instance, his 'Rights and agency' describes the notion of 'goal rights'. A goal right is a sort of right, but it does not work as a side constraint. Instead the satisfaction of a goal right is a sort of good, which may be weighed against other goods.

More important for my purposes in this study is the idea of *agent-relative* good, which Sen advances in the same paper. It is often assumed that good must be a neutral concept: and never relative to a person. That is to say, there is no such thing as good for a particular agent, which might differ from good for another agent. For this reason teleology is often assumed to exclude agent relativity.[32] But consider an ethical theory that says the present generation should give more weight in its decision making to its own wellbeing than to the wellbeing of later generations. This generation, then, should maximize a discounted sum of wellbeing, discounting the wellbeing of later generations. The discounted sum is the generation's objective. Since the ethical theory says the present generation should maximize this objective, the discounted sum is what the theory takes to be good from the point of view of this generation. A different sum might be good from the point of view of another generation. The theory may thus find itself committed to generation-relative good – a particular sort of agent-relative good. If good could not be agent relative, therefore, teleology might rule out discounting.[33] Since, in this study, I clearly cannot rule out discounting from the start, I could not then adopt teleology. But, with Sen's support, I can allow agent-relative good within teleology. I shall consider it in detail in Section 3.4.

Teleology can easily give value to equality. Equality – for instance, equality between generations – can be taken as a sort of good to be included in the objective to be maximized. More specifically, fairness between generations can be included as a good.[34] These are aspects of justice. In taking a teleological line, then, I am not ignoring the demands of justice. I am simply setting aside those particular views about justice that I described in Section 2.1.

To support my teleological approach properly would, naturally, require a great deal more argument than this. It would require a discussion of motivation, for one thing. Suppose that, on balance, it is better for us to reduce our emissions of carbon dioxide, because the benefit to future generations outweighs our loss. How does this fact give us a reason to reduce our emissions, despite the loss to ourselves, and how will recognizing this fact actually bring us to do so? That is the problem of motivation. What motivation have we to do what we recognize as good? One attraction of the idea of justice as mutual advantage is that it is supposed to explain how people can be motivated to act justly: it is to their advantage to do so.[35] I believe that recognizing an action to be good is enough to motivate a person to do it. But I cannot try to participate in this ancient argument here.[36]

A final note. I shall often use the term 'wellbeing' as a synonym for a person's good. This is simply because it may be a more familiar term. My notion of good is wider than Sen's notion of welfare, so wellbeing must not be confused with welfare.

2.3 Distributions of wellbeing

Having laid claim to my right to work in terms of good and bad, rather than justice and rights or something else, I shall now lay down in more detail the framework I intend to work with.

Outlooks

Suppose we are faced with some decision. Suppose, for instance, we are wondering whether to impose a carbon tax, and at what level. We have to choose amongst a range of alternatives: each alternative imposes a tax at some level (one imposes it at a zero level). Consider one of these alternative actions. The result of this action will be uncertain; it depends on all the other things that will happen independently in the world. Let us model this uncertainty by supposing that there is a range of 'states of nature'. One of these states will occur, but we do not know which. Each state determines all the other things that will happen apart from our action. So, given a particular state, the result of our action is determinate.

Take together, then, an action and a state. The two together will determine how the world will progress. They create an 'outlook', as I shall call it. An outlook determines the life path of each person who ever lives. It determines how long each person now living will live for. It determines

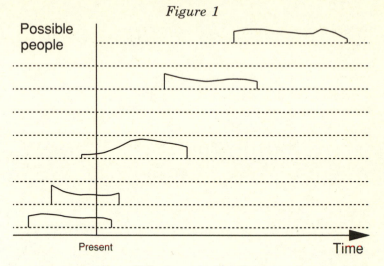

Figure 1

how many people will be born in the future, when each one is born, who she will be, and how long she will live. It also determines how well off each person will be at each moment of her life.

All these features of an outlook can be pictured in a diagram like Figure 1. Figure 1 shows a *distribution of wellbeing*. The horizontal axis represents time. Along the vertical axis are marked all the possible people who might be born. Each one is given a horizontal dashed line. Only some of these people will actually be born. For each one of these, a small graph in the diagram marks the path of her wellbeing, from the time she is born to the time she dies. For convenience, I shall work with a discrete-time version of the distribution, such as Figure 2.

The number of possible people is very large, and very few of them will ever be born. For instance, it is possible that a girl who is now ten might have a child by any one of many million men, though actually she will probably have children by only one. And even by this one man, there are very many children she might have (according to most theories of personal identity, one for each combination of sperm and egg),

Figure 2

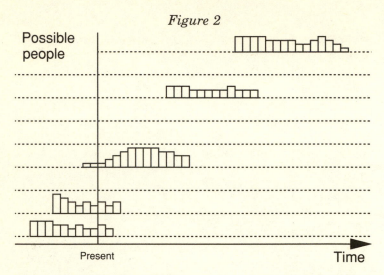

though actually she will have only a particular few. So in practice, most of the vertical dimension of the distribution will contain empty space.

The value function

Let the wellbeing of a possible person i at a time t be g_t^i. More precisely: if there is a person i living at t, let g_t^i be her wellbeing; if there is no such person, let g_t^i have a non-numerical value representing nonexistence, say Ω. So the value of g_t^i is either Ω or a real number.

I shall assume that a distribution of wellbeing contains all the information necessary to determine how good the outlook is. That is to say, the goodness of an outlook depends only on who is born and when, how long each person lives for, and how well off she is at every moment of her life. To put it another way, the goodness of an outlook is a function of all the g_t^is:

$$(2.3.1) \qquad g = g(g_1^1, g_1^2, \ldots g_2^1, g_2^2, \ldots g_3^1, \ldots).$$

I shall call g the *value function*. Since the rest of this study depends heavily on assumption (2.3.1), I need to justify it.

First, I assume that the wellbeing of a person depends only on her wellbeing at all the individual times in her life. To put it in symbols, we can assign each possible person i a wellbeing g^i, which is a function of $g_1^i, g_2^i, \ldots$:

(2.3.2) $g^i = g^i(g_1^i, g_2^i, \ldots).$

If there is never an actual person i, then all the arguments of this function will have the value Ω. In that case, let g^i have the value Ω as well. But if there is ever a person i, then g^i will have a numerical value, which will depend on those values of $g_1^i, g_2^i, \ldots$ that are numerical. These are the values of i's wellbeing during her life. I call this 'the principle of temporal good'. It says, roughly, that all the things that are good or bad for a person must have a date; there are no undated goods. A version of this principle is discussed in my *Weighing Goods*, Chapter 11, and I shall not consider it in detail here. That chapter describes some serious reasons for doubting the principle of temporal good. The most serious, however, applies only to a more complex version of the principle, which involves uncertainty. The simpler version contained in (2.3.2) is less doubtful, and I think that in this context the principle of temporal good is an acceptable assumption, at least as an approximation for the sake of making progress.

Second, I assume that the overall goodness of an outlook depends only on how well off the people are, or more exactly, on which people exist in that outlook, and how well off those people are:

(2.3.3) $g = f(g^1, g^2, \ldots).$

Many of the arguments of this function will take on the value Ω. The value of g will be numerical, though, and it will be determined by those arguments that have a numerical value. I call (2.3.3) 'the principle of personal good'. It says, roughly, that all good belongs to people. A version of it is discussed in *Weighing Goods*, Chapters 8 and 9. Those chapters consider reasons there might be for rejecting it

48 Justice and Wellbeing

(particularly arguments from the direction of egalitarianism), and answer those arguments. So I feel secure in assuming this principle.

The principles of temporal and personal good together give us (2.3.1). The goodness of an outlook is determined by the distribution of wellbeing. But how, exactly? What is the form of the value function g? How is wellbeing *aggregated* across the dimensions of time and people, to determine the overall goodness of an outlook? That is the question I have set myself in this study.[37] I shall not be able to answer it, but I shall survey and assess some views about it. I believe that future research on the costs of global warming should concentrate on answering this question.

Uncertainty

This question is only a part of a more complex one. We started this section wondering, when there is a decision to be made, how to compare the merits of the alternative actions available. Each action will lead, not to a particular outlook for sure, but to many different possible outlooks, one for each state of nature. To evaluate the action, we need to do more than simply evaluate an outlook. We need to evaluate a *prospect*. A prospect is a portfolio of outlooks, one for each state of nature. I have drawn the distribution of wellbeing in just two dimensions: with a dimension for people and a dimension for time. But the distribution is really three dimensional; states of nature constitute a third dimension. And the full question I need to face up to is how good is aggregated across a three-dimensional distribution to determine the goodness of an action.

I believe that expected utility theory describes how good is aggregated across this third dimension of states of nature. That means we can strip off the two-dimensional problem and treat it on its own. First, we can evaluate outlooks, and then rely on expected utility theory to determine the value of actions under uncertainty. That is why I shall be concentrating on the two-dimensional question in this study. But

it is a major limitation of my work. Treating several dimensions together at the same time is a very fruitful means of
investigating the aggregation problem. It reveals surprising
and useful connections between the dimensions. *Weighing
Goods* is devoted to drawing out the conclusions that emerge
from multi-dimensional aggregation. Nevertheless, I believe
that the two dimensions are enough for me to deal with in
this study. If more illumination can be drawn from all three
dimensions together, that is work for the future.

Conditions on the value function

We want to know, then, the form of the value function g.
How is good aggregated across a distribution? We need to
find what conditions or principles govern this aggregation.

   Possible conditions may be divided roughly into two sorts.
There are, first, high-level conditions. These purport to
determine everything about the aggregation – to specify the
function g completely. An example is what I shall call
'complete utilitarianism'. Complete utilitarianism says that
the goodness of an outlook is the sum of the wellbeing of
each person at each time. That is:

$$(2.3.4) \qquad g = \sum_{\{i,\, t \,|\, g_t^i \neq \Omega\}} g_t^i$$

   On the other hand, there are low-level conditions, which
constrain the function g but do not fully determine it. The
principles of personal and temporal good, expressed in
(2.3.2) and (2.3.3), are conditions of this sort.

   In the rest of this study, I shall consider various putative
conditions that might be imposed on the value function, and
discuss how acceptable they are. I shall start, in the next
chapter, with a range of conditions that constitute the bread
and butter of the discussion, in economics, of our responsibilities to the future. These are conditions that come under
the heading of 'discounting'.

Notes

1. 'Circumstances of justice and future generations'; also *Theories of Justice*, pp. 189–203.
2. *A Theory of Justice*, pp. 126–30.
3. *An Enquiry Concerning the Principles of Morals*, Section 3, Part 2.
4. Richards is more explicit in his use of the device for this purpose, in 'Contractarian theory, intergenerational justice, and energy policy'.
5. *A Theory of Justice*, pp. 284–93.
6. *Theories of Justice*, pp. 202–3.
7. *Morals by Agreement*, pp. 204–5.
8. pp. 302–5.
9. For instance, Barry, 'Intergenerational justice in energy policy'.
10. *Morals by Agreement*, pp. 223–32.
11. p. 299.
12. 'Is there a generational contract?'
13. Barry, *Theories of Justice*.
14. *Anarchy, State and Utopia*, pp. 28–33.
15. *Reasons and Persons*, p. 359. Though Parfit is responsible for having brought the nonidentity problem to people's notice, on pp. 364–6 of *Reasons and Persons*, he treats the appeal to rights slightly more sympathetically than I have. Suppose we continue to pollute. Parfit is willing to grant that the people who will exist as a result may have a right to an unpolluted atmosphere. This is a right that, in the nature of things, cannot be fulfilled, because if the atmosphere was unpolluted, these people would not exist. Nevertheless, it could be an objection to polluting the atmosphere that it brings into existence people who have a right that cannot be fulfilled. This could be a reason to control pollution. Parfit seems willing to accept this suggestion, which he ascribes to Tooley in *Abortion and Infanticide*. But if this is indeed a reason to control pollution, it is not the reason that an appeal to rights normally points to. A normal appeal to rights would say that we ought to control pollution because we owe it as a right to someone to do so. And we have seen that actually that is not so. So we need an explanation of how this new type of reason is supposed to work.
16. 'Obligations to posterity', p. 3.
17. 'A moral requirement for energy policy'.
18. An important source for resourcism is Dworkin, 'What is equality?'.
19. See my *Weighing Goods*, pp. 197–8.
20. *Intergenerational justice in energy policy*, p. 20.
21. p. 19.
22. *A Theory of Justice*, pp. 75–83.
23. *Sustainable Development*, pp. 1–22. The idea of sustainable development is a broad one, and many versions of it are neither resourcist nor rigidly egalitarian. See Pezzey, 'Sustainability, intergenerational equity, and environmental policy' for an example.

24. *A Theory of Justice*, p. 291. It is worth mentioning that, in *Theories of Justice*, pp. 197–8, Barry himself recognizes that Rawls's departure from the difference principle in this case is a major concession to the idea of weighing one person's good against another's.

25. 'Intergenerational justice in energy policy', p. 21.

26. In my 'Fairness'.

27. This way is developed in my *Weighing Goods*, pp. 192–200.

28. This is the way of giving value to equality recommended by Atkinson and Stiglitz, *Lectures on Public Economics*, pp. 339–40. It is explained in *Weighing Goods*, Chapter 9.

29. There is some discussion, actually, in an early draft of 'Intergenerational inequality' by Temkin, but I believe this part of the draft is not to be published.

30. There is a much fuller discussion in my *Weighing Goods*, Chapter 1.

31. 'Utilitarianism and welfarism'.

32. Scheffler takes this point of view in *The Rejection of Consequentialism*, and so do many other authors. See, for instance, most of the papers in Scheffler's collection, *Consequentialism and Its Critics*.

33. But notice that discounting could actually be agent neutral. See p. 93.

34. See *Weighing Goods*, pp. 192–200.

35. This, particularly, is why Gauthier recommends the theory in *Morals by Agreement*.

36. The most recent round of this debate was started by Nagel's *The Possibility of Altruism*. For a more recent contribution on the other side, see Smith, 'The Humean theory of motivation'.

37. A technical note. We really only need an *ordering* of outlooks, if we ignore uncertainty. The value function *represents* the ordering. And actually, the ordering may be such that it cannot be represented by a function. This will be so if time continues infinitely; see p. 104. So, technically, we could limit our interest to the ordering rather than the function. But, for convenience, in this study I shall concentrate on the function.

Chapter 3
Discounting for Time

One central question about the value function is how it puts together goods that come at different dates. In cost-benefit analysis and elsewhere, future goods are sometimes counted for less than present goods, and more distant future goods for less than goods in the nearer future. This practice is known as *discounting*. In deciding whether to build a power station, for instance, the electricity the station will produce late in its life is discounted compared with the electricity it will produce earlier. Whether discounting is justified and, if it is, what the rate of discount should be, are hotly debated questions. This chapter surveys the debate.

Much of the argument about discounting is at cross purposes; many of the protagonists are arguing about different things. It all depends on precisely *what* is to be discounted. Some authors are interested in discounting amounts of wellbeing; others in discounting economic commodities such as electricity. It may be that the discount rate should be zero for wellbeing, but positive for commodities. Section 3.1 explains the idea of discounting wellbeing, and separates it from the discounting of commodities. It explains that discounting of commodities is a sort of short cut to the discounting of wellbeing.

Sections 3.2 and 3.3 discuss the question of discounting

commodities. There are two rival theories about it. One says commodities should be discounted at the rate of interest that is available to consumers in the market; the other at the rate of interest that faces producers. Once again, it turns out that these theories are not really in conflict; it is a matter of precisely what is to be discounted. Section 3.2 considers discounting at the consumer rate of interest; Section 3.3 at the producer rate. These sections explain that neither of these methods is adequate for the problem of global warming. Since that is so, and since they address the specific concerns of economists, noneconomists might like to skip those two sections.

Section 3.4 therefore goes on to the question of discounting wellbeing, or 'pure' discounting, as I call it. It considers arguments for and against a positive rate of discount for wellbeing.

3.1 Pure discounting

As I said at the end of Chapter 2, I shall be considering conditions that might be imposed on the value function (2.3.1). One possible condition is that the function is *weakly separable* by times. This means it can be written in the form:

$$(3.1.1) \quad g = h(h_1(g_1^{\ 1}, g_1^{\ 2}, \ldots), h_2(g_2^{\ 1}, g_2^{\ 2}, \ldots), \ldots)$$

where the function h is increasing in all its arguments. If the value function has this form, it means that the value of an outlook can be assessed one date at a time. People's wellbeings at each time t are aggregated to determine h_t. Then the h_ts for all times are aggregated. Weak separability implies that people's wellbeings at one time can be aggregated independently of people's wellbeings at any other time.

A stronger condition is that g is *strongly*, or *additively*, *separable*.[1] This means it can be written in the more specific additive form:

(3.1.2) $g = h_1(g_1{}^1, g_1{}^2, \ldots) + h_2(g_2{}^1, g_2{}^2, \ldots) + \ldots$

A stronger condition still is that g can be written in the form:

(3.1.3) $g = g_1(g_1{}^1, g_1{}^2, \ldots) + g_2(g_2{}^1, g_2{}^2, \ldots)/(1+r)$
$\qquad\qquad + g_3(g_3{}^1, g_3{}^2, \ldots)/(1+r)^2 + \ldots$

where each function g_t shows overall good at time t. In this formula, r is called a *discount rate*. Formally, (3.1.3) could simply be *derived* from (3.1.2) by defining each function g_t as $h_t(1+r)^{t-1}$. What makes (3.1.3) a stronger condition than (3.1.2) is the interpretation of each function g_t as overall good at time t.

Equation (3.1.3) incorporates a *constant* discount rate r. Discounting at a constant rate is often called 'exponential discounting'. I have picked a formula with exponential discounting simply for convenience and ease of presentation. Section 3.4 examines an argument why the rate should be nought, and constant for that reason. But if the rate is positive, I know of no reason why it should be constant.[2]

A value function of the form (3.1.3) is commonly assumed in the literature about discounting. But it is not very plausible. Indeed, even the weakly separable form (3.1.1) is not very plausible. It has the effect of denying any value to the length of people's lives. Imagine something happens that shortens people's lives without making their lives any worse whilst they continue. And imagine that people respond by having more children. (Regular disastrous flooding caused by global warming might perhaps have an effect like this.) Both the death rate and the birth rate increase, then. Imagine the two changes cancel each other out, so that the total population of people living at any time is unaltered. And the conditions of life are unaltered, except that lives are shorter. Then the weakly separable formula (3.1.1) says the world is just as good as before. But this is not very plausible. Intuition suggests the world has become worse, because people's lives are shorter.

This point may be made more precise by means of a

Figure 3

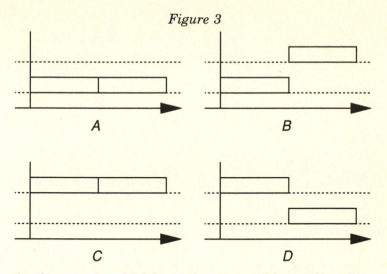

simple example, which has only two possible people and two times. Compare the four alternative outlooks shown in Figure 3. In each, there are two possible people. In alternatives A and C, one person lives for two periods, and the other person does not live at all. In B and D, each person lives for one period. The weakly separable formula gives the goodness of the alternatives as:

$$g(A) = h(h_1(1, \Omega), h_2(1, \Omega))$$
$$g(B) = h(h_1(1, \Omega), h_2(\Omega, 1))$$
$$g(C) = h(h_1(\Omega, 1), h_2(\Omega, 1))$$
$$g(D) = h(h_1(\Omega, 1), h_2(1, \Omega))$$

where h is an increasing function. It follows that if $g(A) > g(B)$, then $g(D) > g(C)$. So if A is better than B, then D is better than C. But it is plausible that A is better than B, and C better than D, because one long life seems better than two short lives.

Cohorts

Weak separability by times, then, is not very plausible, once one takes account of the lengths of people's lives. But it is taken for granted in most of the arguments about discounting, and there is a reason for that. Discounting is a matter of valuing wellbeing that comes at one time differently from wellbeing that comes at another time. We cannot do that unless we have somehow pinned wellbeing down to a time. And that is, effectively, the function of weak separability. It locates wellbeing at a particular time.

If we are to allow discounting (at a positive rate), we shall have to pin wellbeing down to a time, and that means we shall have to assume some sort of weak separability across time. But there are alternatives to (3.1.1). One alternative assumes the value function is weakly separable by *cohorts*. That is to say, it takes the form:

$$g = h(h_1', h_2', \ldots)$$

where h_t' is a function of the good of the people making up the tth cohort. (The tth cohort consists of all the people born at time t.) More particularly, the function might have the discounted form:

(3.1.4) $$g = g_1' + g_2'/(1+r) + g_3'/(1+r)^2 \ldots$$

where g_t' is the good of the tth cohort.

Equation (3.1.4) is not subject to the objection I raised against (3.1.1). It also has the advantage of distinguishing two different questions about discounting. One is the question of how good that comes to a person later in life weighs against good that comes to her earlier. The other is the question of how the good of people who live later weighs against the good of people who live earlier. Since these questions might, perhaps, have different answers, it is useful to separate them.[3] In (3.1.4), the first question is about the determination of each g_t', and the second about the discount rate r.

On the other hand, (3.1.4) is subject to some objections of

its own. The most serious will appear in Chapter 4,[4] but I shall mention two here. One is that I see no reason why the wellbeing of a person should be tied down to the date when she is born rather than, say, the date when she dies. Another is to do with inequality. Imagine a society in which people's lives have ups and downs, although they all work out equally good on balance over a lifetime. At any moment, this society will contain some inequality, but there will be no lifetime inequality between the members of any cohort. One might think that inequality at a time is a bad thing, even if lives as a whole are all equally good.[5] If that is so, this is a value that cannot be captured in a value function of the form (3.1.4).

Pure project evaluation

So neither (3.1.3) nor (3.1.4) is perfectly satisfactory, and I know of no perfectly satisfactory way of pinning wellbeing down to a time. This might be used as an argument against a positive rate of discount for wellbeing: it requires wellbeing to be pinned down in a way that is impossible. But I wish to give the idea of discounting a fair run. So I shall ignore this difficulty, and conduct my discussion in general terms, without insisting on any particular weakly separable formula. I shall ask what the value of the discount rate r is, either in (3.1.3) or (3.1.4), or perhaps in any other formula that separates good by times. Let us call r the *pure* discount rate. I shall be particularly concerned with whether this rate should be nought or positive. A positive rate implies that wellbeing coming later counts for less in the overall goodness of an outlook than wellbeing coming sooner. A rate of nought makes all times count the same; it is impartial between times.

Equations (3.1.3) and (3.1.4) offer, in principle, a way of deciding between whatever alternative actions are available. For simplicity, let us ignore uncertainty, so that each of the alternatives will definitely cause the world to unfold in a particular way. Proceed as follows. For each alternative, assess how good the world will be at each date in the future, on the basis of each person's wellbeing at that date. Then discount these amounts back to present at the pure discount rate. The best action is the one that has the greatest discounted value.

Often, the choice is between a particular *project*, on the one hand, and retaining the status quo on the other. If a project is better than the status quo, I shall say it is a *good idea*. To assess whether a project is a good idea, proceed as follows. For each date, assess the difference between how good the world will be at that date if the project is done and how good it will be if the project is not done. This difference is the *benefit* of the project at that date (or the *harm* if the difference is negative). Discount these benefits and harms back to the present at the pure discount rate. The project is a good idea if and only if the result is positive.

I shall call this the *pure method of project evaluation*. It operates on pure benefits and harms, that is, on the ultimate effects of the project on how good the world is, and specifically on how good it is for people. The pure discount rate is applied to pure benefits and harms.

In practice the pure method is difficult. Two particular difficulties need to be mentioned. One is that the economic data available are not in the form of pure benefits and harms. Instead, the data are about such things as people's incomes and their consumptions of various commodities. Translating the data into pure benefits and harms is hazardous.

The second difficulty is that pure project evaluation requires one to work out *all* the effects of a project, including its indirect effects. The indirect effects may be large and complex. Suppose the project involves some public investment. The process of financing it, by taxation or borrowing, may well withdraw funds from private investment. This will reduce private production in the future, and consequently people's future consumption. This is an 'opportunity cost' of the project. It is one of the project's indirect effects, which must be taken into account in a pure project evaluation.

Short cuts

These two difficulties have prompted economists to adopt what I shall call 'short-cut' methods in practical project evaluation. These methods try to skirt round the difficulties. They also, as a side effect, require one to work with a different discount rate from the pure rate. But these methods are not intended to overrule the pure method. They require different discount rates because different things are being discounted. For discounting pure benefits and harms, the pure rate remains the right one. Here is Joseph Stiglitz making clear the role of the short cuts:

> Any project can be viewed as a perturbation of the economy from what it would have been had some other project been undertaken instead. To determine whether the project should be undertaken, we first need to look at the levels of consumption of all commodities by all individuals at all dates under the two different situations. If all individuals are better off with the project than without it, then clearly it should be adopted (if we adopt an individualistic social welfare function). If all individuals are worse off, then clearly it should not be adopted. If some individuals are better off and others are worse off, whether we should adopt it or not depends critically on how we weight the gains and losses of different individuals. Although this is obviously the 'correct' procedure to follow in evaluating projects, it is not a practical one; the problem of benefit-cost analysis is simply whether we can find reasonable shortcuts. In particular, we are presumed to have good information concerning the direct costs and benefits of a project, that is, its inputs and outputs. The question is, is there any simple way of relating the total effects, that is, the total changes in the vectors of consumption, to the direct effects?[6]

I shall examine the short-cut methods in Sections 3.2 and 3.3. These methods are exclusively concerned with evaluating projects. So in these sections I shall stick to that application. I am ultimately interested in discounting in all its applications, including, for instance, the question of what level of carbon tax to impose.[7] But Sections 3.2 and 3.3 set other applications aside.[8]

The great advantage of the short cuts is that they take their discount rates straight from the market. They discount future commodities at the market rate of interest available either to consumers or producers. They can immediately be used as practical tools for cost-benefit analysis, and to use them there is no need to settle in advance on a pure discount rate for wellbeing. Consequently, these short cuts avoid the need for philosophical argument about what this pure rate should be. Unfortunately, however, my conclusion in the next two sections will be that the short cuts are not available in the special context of global warming. I have nothing against them in principle, but at present I see no alternative to the pure method of project evaluation in this context.

Accordingly, in Section 3.4, I shall come to consider the pure discount rate, which is required for the pure method. This is where philosophical argument is required; it cannot be circumvented. The pure method is far from a practical tool at present. We can hope that, when it comes to detailed quantitative decision making and evaluating particular projects, appropriate new short cuts will in time be found. But we need to have a pure rate first; we cannot avoid the necessary foundational work.

3.2 *Discounting at the consumer interest rate*

In a market economy, each commodity has a price at which consumers can buy or sell it. This is called the commodity's 'consumer price'. It may differ from its 'producer price' because of taxes. If there is an income tax, for instance, producers pay a higher price to buy labour than consumers receive when they sell labour. So the producer price of labour is above the consumer price of labour.

Each consumer decides how much of each commodity she will buy or sell. Consumer theory tells us she will do so in such a way that her marginal rate of substitution between each pair of commodities is equal to the ratio of the consumer prices of these commodities.

Suppose the government is considering a small project. This project will withdraw some resources from the economy, but in return it will produce some commodities as output. For instance, it may use up labour and produce fertilizer. So if the project is done, consumers will have to give up some commodities, but they will receive other commodities in return. Suppose the project is profitable when evaluated at consumer prices. That is to say: the commodities it produces are worth more, at consumer prices, than the commodities it uses up. Then the project could, in principle, be operated as follows. A number of consumers could be required to give up some commodities to provide the resources the project needs. But each of them could be compensated by receiving a bundle of commodities – the output of the project – to a greater value than the commodities she gives up, measured in terms of consumer prices. We know this is possible, because the project is profitable at consumer prices: its output, that is to say, has a greater value than its input. But consumer prices reflect each consumer's marginal rates of substitution between different commodities. Therefore, receiving a bundle of commodities that is worth more in terms of consumer prices makes a consumer better off.[9] So each of these consumers is benefited by the project, provided it is run in the way I have described. No one is harmed by the project, either. If a project benefits some people and harms no one, it is no doubt better to do it than leave it undone. So the project, operated this way, is a good idea.

This piece of economic theory provides a rationale for a particular method of project evaluation. The method is this. List all the commodities that the project uses up, and all the commodities that it produces. Then evaluate them all at consumer prices. The project is a supposed to be a good idea if it is profitable at those prices. Let us call this the *consumer-price method* of project evaluation.

The passage of time comes into this theory as follows. A commodity, as we normally think of a commodity (petrol or doughnuts, for instance), does not have a single price, but a whole series of prices, one for each date. The theory defines the price for a commodity at some date in the future as the price you would have to pay now to buy a unit of that commodity delivered at the future date. In practice, the way you buy future commodities is generally this. You put some money in a bank, where it earns interest. When the date arrives that you want the commodity, you take the money out of the bank and buy it. How much of it you will get depends on how much the price has changed in the mean-time. But, because of the interest the money earns while waiting, £1 used this way will generally buy a greater quantity of a future commodity than it would buy of the same commodity in the present. That is to say, a future commodity is generally cheaper than its present counter-part. Prices decline through time, that is to say. The rate at which a commodity's price declines is called the commodity's 'own rate of interest'.

A typical project uses up resources in the near future, and produces output in the further future. According to the consumer-price method of project evaluation, all the these inputs and outputs should be evaluated at their consumer prices. In this calculation, a commodity in the future will generally have a smaller value than the same commodity in the present, because its future price is less. So this method values a commodity in the future for less than it values a present commodity. In effect, it discounts each commodity. The discount rate is the rate at which the price of the commodity is falling. It is, that is to say, the commodity's own rate of interest.

If the relative price of all commodities remained constant through time, each commodity would decline in price at the same rate. All commodities would have the same own rate of interest, then. Conversely, if commodities change their relative prices over time, that means their own rates of interest differ. In practice, however, changes in relative prices are rarely taken into account in project evaluation. All the commodities that are used and produced by the project are evaluated at their *present* consumer prices. Those used or produced at a particular date, valued at their present prices, are lumped together to produce a money value for the project's 'cost' or 'benefit' at that date. Finally, all these values are discounted at the same rate, to produce an overall value for the project.

The rate used ought obviously to be an approximation to the own rates of interest of all the commodities. In practice, the rate used is the 'real' interest rate available to consumers, in so far as it can be accurately estimated. This is the nominal interest rate less the rate of inflation in the consumer price index. The consumer price index is based on the prices of a particular bundle of commodities. The final effect of using the real interest rate for discounting is that all commodities are discounted at a rate that is actually the own rate of interest of this bundle. Let us call this the *consumer interest rate*.

This consumer-price method, then, in practice goes as follows. First list all the commodities that the project will use up and produce at each date. Evaluate all these commodities at their present prices. Finally, discount these values, according to the date when they occur, at the consumer interest rate. The project is supposed to be a good idea if the discounted total is positive.

This is a considerable short cut compared with the pure method of project evaluation described on page 57. It deals only with marketed commodities, and takes their values from the market. There is no need to work out pure benefits and harms – that is, the effects of these commodities on people's wellbeing. Nor is there any need to establish a pure discount rate; the consumer interest rate, derived from the market, is enough. A further advantage of this theory is that it makes no assumptions about the form of the value function g. It does not even assume weak separability between times; that was not required by the argument.

On the other hand, the consumer-price method is still

faced with one of the difficulties mentioned on page 58. It requires *all* the project's effects on commodities to be accounted for, including its indirect effects. If, say, a government project causes private investment to be reduced, then the resulting future reduction in private output must be treated as a cost of the project. Working out the indirect effects may be difficult. The method of project evaluation mentioned in Section 3.3 goes further, and tries to overcome this difficulty too.

For several reasons, the consumer interest rate will normally differ from the pure discount rate. Firstly, the pure discount rate will differ from the rate at which consumers discount their own wellbeing. Reasons for this are mentioned on pages 71–77 below. Secondly, the rate at which consumers discount their own wellbeing will differ from the consumer interest rate. The reason for this second point is that consumers will not be consuming the same amount of commodities at each time, and we can expect commodities to have diminishing marginal benefit. This is best explained using a simple example.

Suppose there are only two times, and only one commodity, called 'consumption'. Consider the situation of a single consumer, i. Let her consumption at the two times be c_1^i and

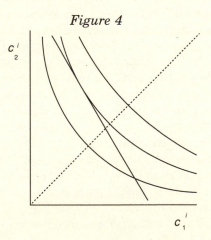

Figure 4

c_2^i

c_1^i

$c_2{}^i$. Suppose the benefits she derives from consumption at the two times are $g^i(c_1{}^i)$ and $g^i(c_2{}^i)$. Suppose consumption brings diminishing marginal benefit, so that g^i is a concave function. And suppose she discounts her future wellbeing at a rate r^i: that is to say, she maximizes the discounted total $g^i(c_1{}^i) + g^i(c_2{}^i)/(1+r^i)$. Let us call r^i the 'consumer's pure discount rate'. This consumer's indifference curves are shown in Figure 4. They are bowed towards the origin because of the diminishing marginal benefit of consumption. Where an indifference curve crosses the 45° line (so $c_1{}^i = c_2{}^i$), its slope is easily shown to be $-1/(1+r^i)$.

The person can trade consumption at one time for consumption at the other by borrowing or lending at the consumer interest rate. If this rate is ρ, she will have a budget line with slope of $-1/(1+\rho)$. She will choose to position herself by trading at a point where her budget line touches an indifference curve. If ρ happens to be the same as r^i, then the slope of the budget line is the same as the slope of the indifference curves where they cross the 45° line. So the consumer will choose to be on this line, which means that she has the same consumption at both times. But the interest rate ρ will be determined by complex interactions throughout the economy, and it will be a great coincidence if it happens to be the same as r^i. If it is different, then the point of tangency will be elsewhere.

Suppose, for instance, that the person does not discount future benefits at all, so r^i is nought. If ρ is positive, the person will make sure, by borrowing or lending money, that her consumption is greater at time 2 than it is at time 1. Then an extra unit of consumption at time 2 will be worth less to her than an extra unit at time 1, because of the diminishing marginal benefit of consumption. Although she does not discount her wellbeing, nevertheless her future consumption is worth less to her at the margin than her present consumption. To generalize beyond the two-period example: if a person does not discount her future benefits, but faces a positive rate of interest, she will make sure her consumption is steadily increasing over time. That way, she can make sure the marginal value of her consumption decreases at a rate equal to the interest rate, and so she will be in equilibrium.

More generally, if a consumer *does* discount her future benefits, she will adjust her consumption plans to make sure that the consumer rate of interest is equal to her pure discount rate *minus* the rate of increase in the marginal benefit she derives from consumption. If her consumption is growing, her marginal benefit is actually decreasing, so the consumer rate of interest will be above the consumer's pure discount rate. It will be equal to her pure discount rate *plus* the rate at which her marginal benefit is declining.

This is a significant conclusion. It is often extended to the economy as a whole, by the simplifying device of assuming the economy consists of one 'representative consumer'.[10] If the economy is growing, the representative consumer's consumption will be increasing over time. Therefore her marginal benefit will be decreasing. The consumer interest rate will be equal to the pure rate of discount plus the rate at which marginal benefit is decreasing.

In using the consumer interest rate, then, the consumer-price method of project evaluation discounts at a rate that differs from the pure rate of discount. But what it discounts is the price of commodities, not wellbeing. Discounting commodities at one rate is perfectly consistent with discounting wellbeing at a different rate. In a growing economy, there are two possible reasons for discounting commodities: firstly the pure rate of discount, and secondly the decline in the marginal benefit of consumption. The consumer interest rate will be above the pure rate of discount. Even if the pure rate of discount is nought, the consumer-price method will still discount commodities at a positive rate.

If this method could be used in the context of global warming, it would be a great convenience. It would spare us the trouble of settling on a pure rate of discount; we could use the consumer rate, which is determined in the market. But I do not believe this short cut is available. I have the following seven objections to it.

First objection: the compensation test is unjustified

My first objection is a general one. It is about how the consumer-price method fixes values in general, and not simply about how it fixes a discount rate. This objection applies to the consumer-price method in all applications, and not just in the context of global warming.

The foundation of the consumer-price method is the so-called compensation test. If a project is profitable when evaluated at consumer prices, it follows that it would be possible to conduct the project in such a way that it benefits some people and harms no one. I explained that on page 61, and I explained that this is the rationale for the consumer-price method. However, in practice, no project is ever conducted in this pleasant manner. All projects in practice harm some people. Just because a project is profitable at consumer prices, that does not mean that *actually* it will harm no one. It means that the people who benefit from the project could in principle fully compensate the people who are harmed, and still end up better off. If the compensation was done, then the project would harm no one in the end, and it would benefit some people. For this reason, a project that is profitable at consumer prices is often said to pass the 'compensation test'. If a project passes the compensation test, then it *could* be run in such a manner as to harm nobody. A project that passes this test is claimed by the theory to be a good idea. The ground offered is simply that the gainers could fully compensate the losers, though actually they will not do so.

This, however, is no ground at all. The fact that the gainers could fully compensate the losers is no reason to think the project a good idea. Nor is there any reason to think a project that fails the compensation test is necessarily a bad idea. Since this is a general point, and generally well understood, I shall not labour it here.[11]

Since, in practice, all projects are good for some people and bad for others, the only way of telling whether or not a project is a good idea is to weigh the benefits to some

people against the harms to others. To do this requires one
to make judgements about the size of people's gains and
losses in wellbeing. It also requires a judgement about the
form of the value function. The idea of the short cut was to
avoid the need to make such judgements, but actually it is
unavoidable.

Projects concerned with global warming are likely to
involve large-scale redistribution of wealth across the world.
Nations are unlikely to bear costs in proportion to the
benefits they receive. It is possible that arrangements can be
made for international compensation. But if they cannot,
this objection will be an important one in practice.[12]

Second objection: commodities with constant benefit

My next objection is not to the fundamental theory of the
consumer-price method, but to the way it is put into practice. The theory says that each commodity ought to be
discounted at its own rate of interest, whereas in practice all
commodities are discounted at the same rate. For many
commodities, this may be a reasonable approximation. But
for some it is plainly not.

One example is leisure. Leisure is a marketed commodity,
and that makes it a good example for explaining the theory.
I shall use it for that purpose, but I do not think it is the
most important example in the context of global warming.
I shall mention more important examples later.

At present, technical progress and capital investment is
leading to an increasing real wage in most economies. That
is to say, wages are increasing faster than the prices of most
goods. The wage can be looked at as the consumer price of
leisure: it is the amount of money a person has to give up in
order to take an extra hour of leisure. So the consumer price
of leisure is increasing faster than the prices of other
commodities. Leisure therefore has a lower own rate of
interest than other commodities. It would be a significant
inaccuracy in a project evaluation to lump in the value of
leisure with the value of other commodities, and discount

them all at the same rate.

Adam Smith thought that 'Equal quantities of labour, at all times and places, may be said to be of equal value to the labourer.'[13] He meant that the benefit a person derives from an hour of leisure is constant over time. Suppose, for a moment, that he was right. More precisely, suppose that, when a particular worker, i, is in equilibrium, the marginal benefit she derives from leisure is constant over time. Then the rate at which she discounts her future leisure will simply be the rate at which she discounts future benefits – her own pure discount rate r^i. (To state this more precisely: her marginal rate of substitution between present and future leisure will be $1/(1+r^i)$.) Because the worker is in equilibrium, her own discount rate for leisure will have to be equal to the market's own rate of interest for leisure. If the worker's pure discount rate is nought, then the own rate of interest on leisure must be nought. Leisure should therefore be discounted, even in the short-cut method of project evaluation, at a rate of nought. In general, if Adam Smith was right, it should be discounted at the worker's pure rate of discount for benefit.

Now, Adam Smith was certainly not right about all workers. Leisure can only have one own rate of interest, because its own rate of interest is determined in the market. Consequently, all workers must have the same marginal rate of substitution between present and future leisure; they must all discount their own leisure at the same rate. But different workers will certainly discount their own *benefits* at different rates. These facts cannot be reconciled if, as Smith claimed, all workers derive a marginal benefit from leisure that is constant over time. For some workers at least, their marginal benefits will have to be changing. This is possible provided leisure has a diminishing marginal benefit: the more leisure a worker takes, the less marginal benefit she derives from it. Then a worker who, say, supplies a decreasing amount of labour over time (and so takes an increasing amount of leisure) will have a marginal benefit from leisure that decreases over time.

Nevertheless, one lesson to be learned from the example of leisure is this. If it *were* true that the benefit a person derives from leisure is constant over time, then leisure should be discounted at a rate that is appropriate for pure benefit. It should not be discounted at the consumer interest rate. Nor, in general, should any commodity that has a constant benefit at the margin. A different rate of discount for such commodities is required by the theory.

In practice, the value of leisure may not be an important issue in the context of global warming. But there are other important commodities whose benefit can plausibly be assumed constant over time. Amongst them are relief from pain or disease, and living in attractive surroundings. One very important one is the saving of life. Saving the life of, say, a twenty-year-old person today presumably brings very much the same benefit as saving the life of a twenty-year-old next year. So life saving seems to have a constant benefit. It ought therefore to be discounted at the pure discount rate, not the consumer interest rate. Both the UK and US governments regularly discount life saving at the same rate as other commodities.[14] There is absolutely no justification for this, even within the theory of project evaluation they are applying.

For many projects aimed at mitigating global warming and its effects, life saving and other constant-benefit commodities, will be important. For them, we need to use a pure rate of discount. So we cannot take the short cut of discounting at the consumer interest rate.

Third objection: market rates do not stretch far enough

The consumer-price method provides a reason for discounting future commodities at the market interest rate available to consumers. Put briefly, the reason is that consumers should be in equilibrium at the market rate, so this rate will be a proper measure of their marginal rate of substitution between present and future commodities.

This may be plausible for short periods, but not for the

many decades that are relevant for global warming. If we had reason to expect steady economic growth to continue throughout the period, we might reasonably reckon on a roughly constant interest rate. But the prospect of global warming itself, together with the depletion of natural resources and the accumulation of other environmental damage, makes steady growth doubtful. We have no good reason to expect the interest rate to stay constant for such a long time. So we have no grounds for using the present short-term rate in very long-term projects. On the other hand, the market does not supply very long-term rates that might be used instead.

In theory, people will have adjusted their marginal rates of substitution to their expectations of distant future interest rates. Take, for instance, a person who is saving now for consumption fifty years in the future. Her marginal rate of substitution between consumption now and consumption fifty years ahead should be determined by the interest she expects to obtain on her present saving, compounded for fifty years. But it is not really credible that this expectation will give us a sound basis for valuing consumption fifty years ahead. Imagine, say, that climatic change is going to cause the world economy to start contracting instead of growing in a few decades time. Consumption will start shrinking. This means its marginal benefit will start to increase with time. Beyond that point it may well be right to use a negative rate of discount for consumption in cost-benefit analysis, and indeed, when the time arrives, the consumer interest rate may well be negative. But it is not plausible that any of this will be reflected in people's present expectations of interest rates.

Fourth objection: future people do not participate

The theory I have described assumes a fixed number of people participating in the market. The value it assigns to future commodities is the value these people assign to them. In practice, the people who participate in the market are

those now living and already grown up. Children and people
not yet born are not included. The values of future commodities to these people are not represented in the market
interest rate.

But it is people who are now children and people who are
not yet born who will reap most of the benefits of any
project that mitigates the effects of global warming. Most of
the benefits of such a project will therefore be ignored by the
consumer-price method of project evaluation. It follows that
this method is quite useless for assessing such long-term
projects. This is my main reason for rejecting it.

I must not overstate this point. Many of the present
population of adults care about their children, their children's children, and other future people. They do, therefore,
assign a value to future commodities because they will
benefit future people. By this one route, benefits to future
people are reflected in the market interest rate. However, I
do not believe that presently existing people are generally
perfectly altruistic. A typical present person, I assume,
values benefits to herself more than she values benefits to
other people. Her marginal rate of substitution between a
present commodity that benefits herself and a future
commodity that will benefit somebody else does not reflect
the ratio of the actual benefits these commodities will bring.
It follows, once again, that the market interest rate does not
measure the true value of future commodities.

In response to this objection, it is sometimes said that the
role of a government, in making decisions about investment,
is to carry out the wishes of its electorate. Stephen Marglin
makes this claim:

> I want the government's social welfare function to reflect only
> the preferences of present individuals. Whatever else demo-
> cratic theory may or may not imply, I consider it axiomatic
> that a democratic government reflects only the preferences of
> the individuals who are presently members of the body
> politic.[15]

Marglin is disagreeing with A. C. Pigou, who says:

There is wide agreement that the State should protect the interests of the future *in some degree* against the effects of our irrational discounting and of our preference for ourselves over our descendants. The whole movement for 'conservation' in the United States is based on this conviction. It is the clear duty of Government, which is the trustee for unborn generations as well as for its present citizens, to watch over, and, if need be, by legislative enactment, to defend, the exhaustible natural resources of the country from rash and reckless spoliation.[16]

Marglin and Pigou are arguing over what the job of a government is. Should a government do what the electorate want, or should it do what is, on balance, for the best? As M. F. Feldstein forthrightly posed the question: 'Should the government act in the best interests of the public, or should it do what the public wants?'[17]

Marglin's side of this argument is not consistent with teleology. I explained in Section 2.2 that a teleological theory is a maximizing theory: it says there is some objective that should be maximized. But the aim of doing what the electorate want is not consistent with maximizing anything. This is one of the lessons to be drawn from a series of discoveries that have be made in social choice theory. A series of theorems from Kenneth Arrow's onwards has shown in one way or another that, if a government only fulfils the wishes of the electorate, then it will not be acting in a way that is consistent with a coherent set of 'social preferences'.[18] It cannot, therefore, have an objective that it maximizes.

I have adopted teleology. So I am committed to siding with Pigou against Marglin. But I do not wish to participate in their argument about political theory. I do not wish to try and defend teleology in this context. Instead, I have a way of sidestepping the argument. I am interested in good. In evaluating a project, I am interested in whether it is better to do the project or leave it undone. Project evaluation, as I understand it, is aimed at answering this question. If teleology is right, that is enough to determine whether or not the government ought to do the project. If teleology is wrong, whether or not the government ought to do the project is a separate question: even if the project is a good idea (because, say, it brings benefits to future generations that outweigh the costs to the present generation), it might still turn out that the government ought not to do it (because, say, its duty is to carry out the wishes of the present generation). But this second question does not concern me in this study. I shall concentrate on what is best, and leave political theorists to debate what the government ought to do. The question of what is best is obviously an important one in its own right. It must also be, at the very least, an important consideration in determining what the government ought to do.

The wellbeing of future people is, of course, a part of overall good. Project evaluation, as I understand it, must therefore take future people into account. But the consumer-price method does not do so, fully. This is my main objection to it.

Fifth objection: imprudence

It is often claimed that, in her own decision making, a typical person attaches more weight to her wellbeing in the near future than to her wellbeing in the more distant future. She discounts her future wellbeing, that is to say; her pure discount rate r^i is positive. Let us call this sort of discounting *imprudence*. If a person is imprudent, then her preferences do not indicate her own wellbeing properly. Take a project that (by increasing taxes and investing the proceeds, say) causes people to consume more in the future and less in the present. Suppose this project increases people's wellbeing on balance. Nevertheless, if they are imprudent, the people might be against it. The consumer-price method of project evaluation will be against it, too. Yet it is, on balance, a beneficial project. If people are imprudent, market interest rates will not reflect the true value of future commodities.

Marglin would find this no reason not to use the consumer interest rate in project evaluation. He thinks that project evaluation should be aimed at people's preferences, not their wellbeing. But I disagree with Marglin's view. I take project evaluation to be aimed at finding what is best. For me, therefore, imprudence would indeed be an objection to using the consumer interest rate.

But are people typically imprudent? The opinion that they are has strong authority behind it. Alfred Marshall supported it; so did Pigou.[19] But I do not know what evidence there is for this opinion. The consumer interest rate appears to be positive,[20] and this implies that people at the margin value future consumption less than present consumption. But there are several reasons apart from imprudence why they might do so. One, illustrated in Figure 4 on page 64, is that their consumption might be increasing over time. Another is that people may derive less benefit from consumption as they grow older. A third is that they may be uncertain about receiving future consumption, because they might die first. Clearly, a positive consumer interest rate is, in itself, no evidence of imprudence.[21]

Since I am unsure that people are typically imprudent, I do not attach much weight to this fifth objection.

Sixth objection: the isolation paradox

My sixth objection will only concern people who take Marglin's view of project evaluation. Since I am not amongst this number, it is not really my objection at all. I mention it only for completeness.

Assume, then, that we wish project evaluation to determine what the government ought to do, and that what the government ought to do is what the electorate would want it to do. The electorate, participating in the market, adjust their marginal rates of substitution between present and future commodities to the consumer interest rate. This interest rate therefore indicates the relative values they set

on present and future commodities. This suggests that
people who share Marglin's political theory should favour
discounting at the consumer interest rate. But there is a
special reason why this is not so.

Suppose I save for the future. I give up some present
commodities for the sake of future commodities instead. The
future commodities will not always come exclusively to me.
Indeed, I may well not use up all of my savings by the time
I die, so that some of the future commodities that will result
from my savings will exist only after I am dead. Why should
I give up my own present commodities for future commodities I shall not actually receive? Partly, at least, it is
because I care about the wellbeing of the people who *will*
receive these commodities. So my motive is partly altruistic.
It may be that I care only about my own descendants, or I
may care about posterity more generally. Whichever it is,
this motivation helps to determine my preferences. Since I
am a member of the electorate, project evaluation, as
Marglin conceives it, should take this motivation into
account.

Because of this motivation, however, and other people's
similar motivations, saving for a future generation is partly
a public good. Each person's saving is valued by others.
There are two reasons for this. The first applies even if
everyone cares only for her own descendants. In the nature
of things, not all the benefit of my saving will be received by
my own descendants. Inheritance taxes, amongst other
things, spread them around. So when I save, I benefit other
people's descendants, and those other people (my contempor-
aries) value that. The second reason applies if people have
a wider concern for posterity beyond their own descendants.
In that case, even the part of my saving that goes to my own
descendants is directly valued by other people.

But public goods are always undersupplied by the free
market. The free-rider problem means that people's individual savings will be less than the optimum level: less than the
level individuals would themselves choose if free riding was
prevented. The government can improve the situation by

compulsory saving. It ought to use a lower discount rate than the consumer rate in deciding which projects to undertake. This will lead it to do more investment than individuals would choose to do on their own, and this is a form of forced saving. But this is not to ignore the preferences of its electorate. It is actually to implement their preferences; it overcomes a market failure that prevents them from putting their preferences into effect as individuals.

This argument for using a discount rate lower than the consumer interest rate is know as the *isolation paradox*. It is associated with Stephen Marglin and, particularly, Amartya Sen.[22] Within its own framework, it is certainly valid. I shall mention a significant response to it later.[23]

Seventh objection: global warming is not marginal

The theory behind the consumer-price method applies only to small projects, which will not significantly alter prices. For larger projects, the short cut is not available, at least without major modifications. Projects to do with global warming will not be small.

This objection is quick to state, but no less significant for that. It is one of my major objections to the consumer price method.

3.3 Discounting at the producer interest rate

A public investment project will use up resources. However it is financed, by borrowing or by taxation, these resources will have to be diverted from the private sector. Some may come out of private consumption. But some, at least, are likely to come out of private investment; to some extent at least, public investment will displace private investment. This is an effect that must be taken into account in evaluating a public project. It suggests a new short cut.

Notice, as a preliminary, that the producer interest rate in an economy is normally higher than the consumer rate.

That is to say, when a producer borrows money, the interest it has to pay is higher than the interest a consumer who lends money ultimately receives. The difference results from taxes. Suppose a consumer pays income tax of 25%, say. Then if a producer pays interest at 8%, the consumer receives interest at 6%. To be accurate, I should say that the consumer own rate of interest on any commodity is 25% below the producer own rate of interest on that commodity. But let us suppose for simplicity that the relative prices of commodities remain constant. Then all commodities have the same consumer own rate of interest, 25% below the producer rate.

An economy's productive activity is, in effect, a process that transforms present commodities into future commodities. When producers are in equilibrium, the marginal rate of transformation between present commodities and future commodities is given by the producer interest rate. If the rate is 8%, then 100 units of commodities now can be transformed into 108 in one year's time.

Now, let us consider how to evaluate a public project. For the moment, let us take for granted the consumer-price method of project evaluation. Suppose the consumer interest rate is 6%, and the producer rate 8%. Consider a project that requires investment of 100 units of commodities this year, and will deliver as output 107 units next year.

Imagine first that this project will have no effect on private investment, so the whole 100 units of investment will be taken from present consumption. Consumption will be reduced by 100 units this year, and increased by 107 units next year. Granted the consumer price method, next year's consumption should be discounted at the consumer interest rate, 6%. The present value of the project is therefore $-100 + 107/1.06$. This sum is positive. So the consumer-price method, which we are taking for granted, says the project is a good idea. This is no surprise: its rate of return is 7%, which is more than the consumer interest rate of 6%.

Next imagine the project will draw its resources from private investment. None will come from consumption, so the project does not reduce this year's consumption at all. Next year it will deliver 107 units. However, if the project is not done, private industry will retain the 100 units of investment, and deliver 108 units of consumption next year. The net effect of the project on next year's consumption is therefore −1 unit. The present value of the project, discounted at the consumer interest rate 6%, is $0 + (-1)/1.06$. This is negative. So the consumer-price method says this project is a bad idea. It is a good idea if it is financed out of consumption, but a bad idea if it is financed out of investment.

If a project withdraws investment from the private sector, it has an 'opportunity cost'. Apart from its own direct inputs and outputs, it has indirect effects that must be taken into account. These may turn a project that looks a good idea on the surface into a bad idea. It is easy to see from the example that, if a project displaces private investment, it will be a bad idea unless its direct rate of return is as high as the producer interest rate. To put it another way: to be a good idea a project that displaces private investment must have a positive present value when discounted at the producer interest rate.

This points to a new short cut in project evaluation. This short cut can be used only for projects that draw all their resources out of private investment. Such a project will be a good idea if and only if the present value of its direct inputs and outputs is positive when discounted at the producer interest rate. The short cut, then, is to work out only the *direct* inputs and outputs of the project, ignoring the influence on private investment, and discount them at the producer interest rate.[24] Let us call this the 'producer-interest-rate method' of project evaluation.

The argument about opportunity cost might be generalized to support a full-blown producer-price method. On pages 83–90, I shall consider an argument offered by Peter Diamond and James Mirrlees in favour of a producer-price method. But the rest of the literature I shall be considering concentrates on the interest rate only. Until I come to

Diamond and Mirrlees, I shall do the same. Implicitly I shall suppose that, apart from the interest rate, the relative producer prices of commodities are the same as their relative consumer prices.

I must emphasize that the producer-interest-rate method does not supersede or overrule the consumer-price method. The difference is that the consumer-price method requires all the project's effects to be laid out explicitly, and then discounted at the consumer interest rate. The new short-cut method requires only the direct effects to be laid out. The answer will be exactly the same, but it will have been arrived at with less trouble. The consumer-price method, remember, was itself a short cut. So this is a short cut in a short cut.

I said that the producer-interest-rate method will work only when public investment displaces private investment completely. But actually, this condition is oversimplified. Whether the producer-interest-rate method will give a proper evaluation of a project depends, not just on the project's immediate effect on private investment, but on its effect throughout its entire lifetime.

To see why, take another example. Suppose once again that the consumer interest rate is 6% and the producer rate 8%. Consider a long-term public project that yields a rate of return of 7%, ignoring indirect effects. Suppose it requires as input 100 units of commodities this year. It will yield no output for 40 years, but after that time it will produce 1497 units of commodities. (This is 7% per year, compounded for 40 years.) But suppose its 100 units of investment will be drawn from the private sector, where it can obtain a rate of return of 8%. Suppose, however, that if the 100 units is invested in the private sector, it will yield 8 units of commodities each year, and these will be consumed rather than reinvested. Discounted at the consumer interest rate of 6%, the public project has a net present value of 46 units of commodities. Private investment of 100 units, on the other hand, has a net present value of 33 units. The public project is therefore a good idea, even though it has a rate of return less than the producer interest rate. The reason is that, whereas the private sector will consume its output, the public project in effect forces reinvestment for the full forty years. Discounting at the producer interest rate would give the wrong answer for this project.

This example shows that the question of whether to use the producer-interest-rate method of project evaluation is a complex one. It depends on whether public investment completely displaces private investment during the project's lifetime. If it does, the producer-interest-rate method may be used. If, at the other extreme, it should turn out that a public project will not affect private investment at all, then no short-cut method is needed for evaluating it. The project's direct effects are all its effects, and they should be discounted at the consumer interest rate. If a project *partly* crowds out private investment, then some other short cut may be available. It may be possible to evaluate the project by taking its direct effects only, and discounting them at some other rate. Normally, the appropriate rate will lie somewhere between the consumer interest rate and the producer interest rate.

Most of the literature on the discount rate within economics revolves around the question of what short cut to use in what circumstances.[25] A comprehensive article by Joseph Stiglitz illustrates the complexity of the problem.[26] Stiglitz concludes that different types of project call for different discount rates. Not surprisingly, one important issue in fixing the value is how the project is to be financed. For reasons given on pages 90–92, I do not believe that any short cut derived from the producer interest rate is available for problems to do with global warming. I shall therefore not dwell on this literature. But I shall pick out of it a few points that are relevant to this study.

International mobility of capital

The first is that recent developments in the international capital market have made a difference to the situation.

Robert Lind has argued that the market now works in such a way that any public investment project will effectively be financed by borrowing on the international market.[27] It will not displace domestic private investment. As a result, Lind recommends discounting at the consumer interest rate.

Lind's argument, however, betrays a parochial attitude. He says:

> 'In a world of integrated international capital markets and a high degree of capital mobility, the link between total domestic saving (private saving plus the government surplus) and domestic investment is broken. It is worldwide savings, the sum of all government surpluses and deficits, and total world private investment that are linked.'[28]

This means that US public investment, say, will not displace US private investment. It will therefore not impose an opportunity cost of capital on the US economy. But it *will* displace private investment in the world as a whole. The opportunity cost is exported, that is to say. I said on page 23, however, that the problem of global warming can only be solved by international cooperation. A parochial point of view is pointless. We must be concerned with the global opportunity cost of a project. In the context of global warming, therefore, I reject this particular argument against the producer-interest-rate method.

The isolation paradox

The second point is raised by the isolation paradox. I explained this argument on pages 75–77. If saving is partly done for altruistic motives, then it is partly a public good. Because of the usual free-rider problem for public goods, saving will be undersupplied by the private market. This creates a case for the government to intervene, and in some way artificially increase the level of saving. Or so it seems. One way a government might want to intervene is by evaluating public projects at a lower rate of interest than the private market does. This will lead the government to adopt some projects that the private market rejects. The isolation paradox, then, is an argument for discounting at less than the consumer interest rate.

There is, however, more to be said about the isolation paradox. The point is made in a general form in an article by T. Bergstrom, L. Blume and H. Varian.[29] Consider any public good that private individuals are supplying to some extent. Now suppose the government intervenes to supply the good, attempting to increase the total supply. The effect, according to these authors, will simply be that the private individuals reduce their supply by an amount exactly equal to the government's supply. So there will ultimately be no effect on the total supply of the public good. The argument for this conclusion is simple, but I shall not spell it out here.

Anticipating Bergstrom, Blume and Varian's general conclusion, an earlier article by Peter Warr and Brian Wright applied a similar argument to the isolation paradox.[30] Their conclusion is that, if individuals are saving for altruistic motives, any attempt at extra saving on the part of the government will simply cause them to save less.

This point amounts to a particular reason why public investment will fully displace private investment. Subject to certain limitations that have been pointed out by Amartya Sen and David Newbery,[31] the argument is certainly valid. If public investment fully displaces private investment, then the producer-interest-rate method is an appropriate short cut for project evaluation. Warr and Wright's argument supports this method, therefore. It supplies only a partial defence of it, though, since it applies only to altruistic saving. It is only a negative defence: it says that the isolation paradox is not a successful argument *against* the producer-interest-rate method.

Diamond and Mirrlees

'Optimal taxation and public production' by Peter Diamond and James Mirrlees lays out some fundamental theory about discounting at the producer interest rate. On page 61, I described a general argument for the consumer-price method of project evaluation. Diamond and Mirrlees, on the other hand, present a general argument for a producer-price method. They argue that a project should be accepted if and only if it is profitable when evaluated at producer prices. One implication is that discounting should be at the producer interest rate.

Here is a sketch of their argument. Diamond and Mirrlees assume the government is free to set any rate of tax (or subsidy) it chooses, on any commodity. But these commodity taxes are the only taxes it can impose; in particular, it cannot use lump-sum taxes. If, therefore, it wishes to raise any revenue, or redistribute income amongst the population, it must do so by means of commodity taxes. If a commodity is taxed, that means consumers and producers do not all face the same price for it. Consequently, they will not all have the same marginal rate of substitution between it and other commodities. This implies that the economy is technically inefficient: within its resources, it would be technically possible to make someone better off without making anyone worse off.

Some sort of inefficiency is inevitable in the economy, then. Nevertheless, Diamond and Mirrlees argue that the *production* system should be efficient: the inevitable inefficiency should be located elsewhere. They show that, if the production system is inefficient, then it would always be possible to alter taxes in such a way as to make some people better off without making anyone worse off. If ever production is inefficient, then, there will always be scope for an improvement.[32] I need not spell out the details of how the improvement can be made. But I should mention that the argument does depend on one strong assumption: there must be no pure private profits in the economy. Perfect competition, with free entry and exit, would have this effect. So would constant returns to scale in private production. In one way or another, pure profit must be kept to nought if the Diamond-Mirrlees argument is to work.

How can efficiency in production be achieved? The

production system will be efficient if there is one uniform set of producer prices, and all producers are maximizing their profit at these prices. The private sector, we can assume, will be maximizing its profit at whatever prices it faces. So how should the government decide on what production *it* should undertake? To maintain efficiency, the government must maximize profit at the same prices as the private sector. This means it must never undertake a project that is not profitable at those prices. When it is evaluating any project, therefore, it must do so at the prices faced by private industry – at producer prices, that is. This is the only way to maintain production efficiency. That is the argument for the producer-price method of project evaluation.

Now, the first thing to notice about this argument is that it does not say a project is a *bad* idea if it is *not* profitable at producer prices . Suppose a project comes up for consideration that is profitable at consumer prices, but not at producer prices. On page 61, I presented an argument to show that any project is a good idea if it is profitable at consumer prices. That argument has a weakness over the matter of distribution between gainers and losers, which I described on page 67. But, setting that problem aside, it remains valid now. It tells us that, since the project we are considering is profitable at consumer prices, it will actually be a good idea. The Diamond–Mirrlees argument does not deny that. It only says, if this project is a good idea, then something even better than the project could be done by altering the tax system instead.

Suppose, for instance, some project will earn a rate of return on investment that is lower than the producer rate but higher than the consumer rate. This project may be a good idea. If so, and if nothing else is done, at least this project should go ahead. But something even better could be done. It might well be better to reduce the producer interest rate. It is the tax system that maintains a difference between the consumer interest rate and the producer interest rate. If the tax on interest was reduced, the producer rate would fall. This would encourage more private investment. It would increase investment in a way that maintains efficiency in the productive system. The project we are considering might then become profitable at the new lower producer interest rate. That is to say, it might become acceptable according to the producer-interest-rate method of evaluation. The Diamond-Mirrlees argument does not actually guarantee that this particular change in the tax system would be the right one. But it does guarantee that *some* change would be better (better for some people and worse for nobody) than simply undertaking this one project.

Diamond and Mirrlees's argument for the producer-price method only works if the government uses taxes in the best possible way. If it does not, then it might indeed be a good idea to undertake certain projects that are not profitable at producer prices. Doing so might partly undo the damage done by a poor tax system. I said on page 20 that many of the problems raised by global warming might be better solved by changes in the tax system than by government investment. But, so long as the tax system is imperfect, there will be individual government projects to consider, and they will have to be evaluated in the context of the imperfect tax system. In that case, we cannot rely on the Diamond-Mirrlees argument for producer-price evaluation.

The argument, however, still raises one puzzle that needs to be sorted out. Let us imagine the tax system is the very best possible (constrained by the impossibility of lump-sum taxes). There will still be taxes, naturally, and that means that producer prices will still differ from consumer prices. Consequently, there will still be projects that are profitable at consumer prices but not at producer prices. Are these projects a good idea or not? The Diamond-Mirrlees argument apparently says they are not. So it seems to be a direct attack on the consumer-price method of evaluation. It seems, not simply to be offering the producer-price method as a useful short cut, but actually to be saying the consumer-price method is wrong. Yet, we have the argument on page 61 to support the consumer-price method. How are these arguments to be reconciled?

To answer this question, I shall use a simple example and diagram adapted from Diamond and Mirrlees. The diagram

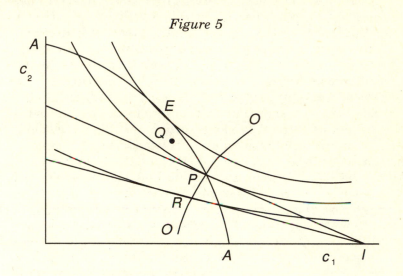

Figure 5

is Figure 5. This example assumes there is only one consumer. Besides making the diagram simple, this assumption has the second incidental benefit of sharpening the apparent conflict between the arguments. It removes the general objection I mentioned on page 67 to the argument on page 61. That objection was to do with distribution between different consumers, and when there is only one consumer it does not apply. When there is only one consumer, the argument in favour of the consumer-price method of project evaluation is hard to fault.

The example also assumes there are only two commodities, say consumption this year and consumption next year. This year's consumption can be converted into next year's consumption through the production system. The government needs some consumption goods for its own purposes; that is why there will have to be taxes. The curve *AA* in Figure 5 is the frontier of the economy's production possibilities, after deducting the government's requirements. Point *I* is the consumer's initial endowment. The government controls the prices that face the consumer. That is to say, it controls the consumer interest rate. Different interest rates will give the consumer different budget lines. Two, *IR* and *IP*, are shown in the diagram. On each possible budget line, the consumer will choose the best point she can for herself, given her preferences. The diagram shows three of her indifference curves. Her choices define her offer curve *ORPO*. The best possible position for this economy is where the offer curve crosses the production frontier at point *P*. This is not the technically most efficient position; that is point *E*. But it is the best point that can be attained without lump-sum taxes. It can be supported by a consumer interest rate given by the slope of the budget line *IP*, and a producer interest rate given by the slope of the production frontier at *P*. (In each case, the slope is $1/(1+\rho)$, where ρ is the interest rate.)

Now consider a small project that would move production from *P* to *Q*. (It would convert a little more of present consumption into future consumption.) This project is profitable at the consumer interest rate, because *Q* is above the consumer's budget line. It is not profitable at the producer interest rate, because *Q* is below the production frontier. So this is a suitable project to test out our apparent conflict of arguments. Is it a good idea or not?

The answer is that it *would be* a good idea, but it is actually not a possible project. It would be a good idea because if it were carried out it would put the consumer on to a higher indifference curve. The fact that it is profitable at the consumer interest rate tells us this directly. That is why the argument for consumer-price evaluation is in fact correct, in this example with a single consumer.

However, this project is not possible because it is not on the consumer's offer curve. It is *technically* possible, because it is within the production frontier. But the offer curve tells us what can practically be achieved in this economy, given the fact that the government has only commodity taxes available. Suppose the government were to try and carry out the project. The project requires the consumer to reduce her present consumption. This she will not do unless her interest rate is raised. But raising the rate will only induce her to move up her offer curve; it will not induce her to move to *Q*. To get her to give up enough present consumption to finance the project, the rate would have to be raised a lot. So she would have to receive large repayments next year – more than the project itself can produce.

Put this another way. The project will have to be financed, and the operation of financing it will have various effects. These are *indirect* effects of the project. A project that moved production from *P* to *Q* would be a good idea. But in practice, a project that looks at first as though it will move production from *P* to *Q* will not actually do that at all. Financing the project will require changes in taxes. The indirect effects of these taxes make a move from *P* to *Q* impossible. Inevitably, taking account of indirect effects, any project can only move the economy to a point on the offer curve, somewhere within the production frontier. Now, the consumer-price method of project evaluation requires one to take account of indirect effects. Taking account of indirect effects, any project must end up on the offer curve below the frontier. All this part of the offer curve is below the consumer's budget line *IP*. Therefore, no project can possibly be profitable at consumer prices, once indirect effects are taken into account.

Diamond and Mirrlees's argument, then, does not say the consumer-price method is wrong. It does say that, if taxes are used properly, the consumer-price method will be a waste of effort. It will have to take account of the indirect effects of financing the project, and those will generally cancel out the direct effects. It is easier to concentrate directly on keeping the production system efficient, and this is best achieved by evaluating projects at producer prices, without worrying about indirect effects.

The upshot, then, is this. Diamond and Mirrlees have no objection to the consumer-price method in principle. But in practice, when taxes are used properly, it is otiose. Their argument favours the producer-price method. But, like the other arguments I have discussed in this section, it favours it ultimately because it is a short cut.

My objections to discounting at the producer interest rate

As I said on page 80, the producer-interest-rate of project evaluation is a short cut in a short cut. It takes the consumer-price method for granted, and tries to abridge it by removing the need to deal with indirect effects. Consequently it is subject to many of the objections I raised to the consumer-price method in Section 3.2.

It is not fully subject to all of them, however. As I said on page 74, my most important objection to the consumer-price method is that future people are not represented in the market. The market will therefore understate the true value of future commodities. But the producer-interest-rate method is supported by considerations of opportunity cost. It is a matter of comparing the project with productive opportunities available in the private sector. This method is appropriate when public investment fully displaces private investment. Under these circumstances, undertaking a public project with a lower rate of return than the producer rate will actually diminish the economy's net production of commodities at *every* time. The example on page 79 shows this. Future people have no say in determining the opportunity cost of investment, but nevertheless it is foolish to embark on a project whose opportunity cost is greater than its benefit.

However, there is still a point in the objection that future people are not represented in the market. The producer interest rate is itself determined by market forces. Suppose future people were to obtain representation in the market somehow. Suppose, for instance, the government started to undertake public investment on a large scale, thereby transferring more resources to the future. This would reduce the producer interest rate. That is to say, it would reduce the opportunity cost of investment. The final result could be a new equilibrium with the government's projects turning out profitable at the new lower producer interest rate. The trouble with the producer-interest-rate method of project evaluation is that it is only useful for marginal projects. On page 77, I raised the same objection to the consumer-price method. This objection carries over to the producer-interest-rate method, and appears here in this particular guise. It prevents the interests of future generations from being properly accounted for. This is a serious objection to the method.

But my main objection is a different one. The producer interest rate on the market does not actually represent the opportunity cost of investment. It represents the private rate of return on investment, and this is different. The difference is the external costs imposed by private investment. Most private investment has the effect of increasing the emission of greenhouse gases into the atmosphere. If these gases are harmful, therefore, private investment imposes an external cost that does not appear in the private rate of return. If the producer-interest-rate method of project evaluation is to be used, there must at the very least be a correction for this external cost. But we cannot make the right correction without knowing in advance the cost imposed by the greenhouse gases. These gases do their damage over a long period of time. Therefore we cannot evaluate their costs without first knowing how to weigh together amounts of damage done at different times. That is to say, we need in advance a discount rate for damage. Evidently, this will need to be a *pure* discount rate, applicable to amounts of actual harm. We shall need to do some pure discounting of harms before we can even consider applying the producer interest-rate method properly.

This single consideration is enough to ensure that the short cut offered by the producer-interest-rate method is not available in planning what to do about global warming. At least, it is not available at this stage. Later, after we have fixed a value on the harm done by greenhouse gases, using a pure discount rate, we may be able to apply an appropri-

ate correction to producer interest rates. But that prospect
is speculative and far in the future.

3.4 The pure discount rate

The pure discount rate, such as r in (3.1.3) on page 54 or
(3.1.4) on page 56, is applied to pure benefits and harms, or
pure goods and bads. This section asks whether it should be
nought or positive. I shall review the arguments on both
sides. If it should turn out that the rate ought to be positive,
the question would then arise of what, precisely, it ought to
be. But I know of no discussion of this point; the literature
only debates whether or not the rate should be nought.

Impartiality

Let us start with Henry Sidgwick:

> It seems . . . clear that the time at which a man exists cannot
> affect the value of his happiness from a universal point of
> view; and that the interests of posterity must concern a
> Utilitarian as much as those of his contemporaries, except in
> so far as the effect of his actions on posterity – and even the
> existence of human beings to be affected – must necessarily
> be more uncertain.[33]

This much is incontrovertible: *from a universal point of view*
the time at which a man lives cannot affect the value of his
happiness. A universal point of view must be impartial
about time, and impartiality about time means that no time
can count differently from any other. In overall good, judged
from a universal point of view, good at one time cannot
count differently from good at another. Nor can the good of
a person born at one time count differently from the good of
a person born at another. Therefore, if either (3.1.3) or
(3.1.4) is to represent overall good judged from a universal
or impartial point of view, the discount rate r must be
nought. The question, though, is whether we should adopt

an impartial point of view.

I must define my terms more accurately. A value function is *impartial* if each time or each cohort counts the same. The value functions (3.1.3) and (3.1.4) are impartial if g is unaffected by permuting the g_ts or g_t's. Plainly, impartiality implies that the pure discount rate is nought. Impartiality is not the same as agent neutrality. A value function is *agent neutral* if every agent has the same function. In this study, I am concerned with one particular sort of agent neutrality: generation neutrality. Conceivably, different generations might have different value functions. If so, the value function is *generation relative*; if not, *generation neutral*.

If a value function is generation neutral, so that every generation accepts it, it is natural to expect it to be impartial between generations. But it need not be. Indeed, one type of nonimpartial function could quite plausibly be generation neutral. If a value function has a constant positive discount rate – exponential discounting, that is, as (3.1.3) and (3.1.4) have – then every generation could quite plausibly accept it. Exponential discounting places every generation in the same position relative to its neighbours: its wellbeing counts in the function a bit less than the previous generation's wellbeing and a bit more than the succeeding generation's. Maximizing a value function with exponential discounting would allow each generation to give itself a little more weight than its successors, and this seems quite a plausible objective for each generation to accept.

It is not *very* plausible, though. A value function with exponential discounting requires each generation to give more weight to the wellbeing of earlier generations than it does to its own. The cohort version of the function, (3.1.4), would require each cohort to value the wellbeing of older people, many of whom are still living, more than it values its own wellbeing. It is a heavy demand to place on anyone to ask her to value someone else's wellbeing more than her own. *Any* version of a value function with positive expo-

nential discounting would require a generation to give
enormous weight to generations who lived in earlier centuries. We probably cannot do much to benefit people who
lived long ago, so this may not seem like an unbearable
burden. But it can be tested with a thought-experiment.
Shakespeare seems to have believed that literary immortal-
ity is a great boon to an author. If he was right, we can still
benefit Shakespeare by propagating his fame. According to
a value function with exponential discounting, benefiting
Shakespeare in this way is a far more important aim than
anything we might do for ourselves. If we find this implausible, we cannot accept such a value function. If there is to be
positive discounting, it is much more plausible to think that
each generation will discount both forwards and backwards
in time; it will count the wellbeing of both earlier and later
generations for less than its own. If it does, the value
function will be generation relative. The conclusion I draw
is that generation neutrality does not strictly imply impartiality, but that a generation neutral value function would
very probably be impartial.

Now we can come back to the question of whether the
value function should be impartial. It is a common opinion
that good must be agent neutral, just because the concept of
good implies agent neutrality.[34] In so far as agent neutrality implies impartiality, it would follow that good must be
impartial.[35] I insisted in Chapter 2 that the value function
g is intended to represent goodness. So it would follow that
the value function must be impartial, and the discount rate
in formulae such as (3.1.3) and (3.1.4) must be nought.

But I do not find this a conclusive argument. First, I am
not convinced that good must be agent neutral.[36] Second,
I have just offered some possible grounds for doubting that
agent neutrality − specifically generation neutrality −
strictly implies impartiality. Consequently, I do not believe
that good is necessarily impartial. At least, I do not think
that the concept of good implies impartiality.

Even if it is, though, that would scarcely be an adequate
argument against discounting. We need to know whether a

generation, in its decision making, is justified in discounting the wellbeing of future generations. Or alternatively, ought each generation to act impartially? Even if good were necessarily impartial, it would not follow that a generation should act impartially. That would only follow if a generation should necessarily do what is best – if it should maximize good, that is. Teleology is the theory that it should, and I have adopted teleology. But I do not wish to dodge the significant arguments about discounting. I must recognize the possibility that teleology may be false.

Because I favour teleology, I think that each generation should do what is best, but I allow for the possibility that what is best may not be impartial. That way, I leave room for the possibility of discounting. Alternatively, one might insist that good is impartial, but allow for the possibility that a generation need not do what is best. That would be a different way to leave room for discounting. In either case, the important question is whether a generation ought, or ought not, to be impartial in its actions. This is the question I shall pursue in this section. I shall discuss the arguments on either side.

Utilitarianism

On the side of impartiality is the utilitarian tradition, represented by Henry Sidgwick in the quotation on page 92. Sidgwick takes it for granted that a utilitarian must aim to maximize good, conceived impartially. The doctrine of impartiality – 'each to count for one, and none for more than one' – lies at the heart of utilitarianism. Indeed, utilitarians often see impartiality as the very essence of morality. Here is an example from economics: John Harsanyi says 'Since Adam Smith, moral philosophers have often pointed out that the moral point of view is essentially the point of view of a *sympathetic* but *impartial* observer.'[37] So, whatever are the arguments that ground utilitarianism, those arguments support impartiality and a zero discount rate. I do not intend to review those arguments here. But I do think they create a strong presumption against discounting. The onus of proof is on those who deny impartiality.

With the weight of utilitarianism behind them, few supporters of a zero discount rate have felt the need to offer an explicit defence. Frank Ramsey, famously, simply announces that discounting is 'a practice which is ethically indefensible and arises merely from the weakness of imagination'.[38] Derek Parfit chooses only to answer arguments against impartiality, rather than provide a positive argument for it.[39] Robert Solow asserts that 'in social decision-making . . . there is no excuse for treating generations unequally',[40] but again he offers little argument. He does say:

> We have actually done quite well at the hands of *our* ancestors. Given how poor they were and how rich we are, they might properly have saved less and consumed more. No doubt they never expected the rise in income per head that has made us so much richer than they ever dreamed was possible.[41]

But this is an exhortation rather than an argument: we have been lucky, so let's be generous. Solow seems to think our ancestors made a mistake, and it hard to know what conclusion we ought to draw from that.

Rawls

John Rawls believes in impartiality. But, not being a utilitarian, he recognizes the need to offer a defence. His argument is brief, and I reproduce it here:

> Since in justice as fairness the principles of justice are not extensions of the principles of rational choice for one person, the argument against time preference must be of another kind. The question is settled by reference to the original position; but once it is seen from this perspective, we reach the same conclusion. There is no reason for the parties to give any weight to mere position in time. They have to choose

a rate of saving for each level of civilization. If they make a distinction between earlier and more remote periods because, say, future states of affairs seem less important now, the present state of affairs will seem less important in the future. Although any decision has to be made now, there is no ground for their using today's discount of the future rather than the future's discount of today. The situation is symmetrical and one case is as arbitrary as the other. Since the persons in the original position take up the standpoint of each period, being subject to the veil of ignorance, this symmetry is clear to them and they will not consent to a principle that weighs nearer periods more or less heavily. Only in this way can they arrive at a consistent agreement from all points of view, for to acknowledge a principle of time preference is to authorize persons differently situated temporally to assess one another's claims by different weights based solely on this contingency.[42]

This argument seems to me mistaken. Rawls confuses impartiality with generation neutrality. Consequently, the last sentence of this quotation is incorrect. A principle of time preference does not necessarily authorize persons differently situated temporally to assess one another's claims by different weights. Take a principle with exponential discounting, which discounts generations at a constant positive rate throughout history. This principle incorporates time preference; it is not impartial. Nevertheless, it could be generation neutral. It would be possible for each generation to accept and act on this principle. Each would be required to give more weight to the good of its predecessors than it does to itself, but it would be permitted to give less weight to its successors.

Indeed, I can think of reasons why this principle might reasonably be accepted by the parties in the original position, at least if history has no beginning and no end, so there is no first generation and no last. Compared with an impartial principle, exponential discounting treats each generation less favourably relative to its predecessors. But in compensation, it treats each more favourably relative to its successors. And it has the advantage, compared with an impartial principle, of putting less strain on each generation's self-control. On page 93, I gave reasons why a generation might not accept a value function with exponential discounting as its *measure of good*. But it seems more plausible the exponential discounting might be a principle that the parties in the original position would agree *to act upon*.

I cannot, therefore, accept this argument of Rawls's.

The principle of personal good

We are concerned with impartiality across time. But a value function may fail to be impartial in other ways too. For instance, many people think we should give more weight in our decision making to our neighbours and compatriots, than to strangers and foreigners. This idea implies a value function that is not impartial between people; it discounts for distance.

The principle of personal good, which I expressed in (2.3.3) on page 47, is inconsistent with any value function that is not impartial, and gives different weights to the wellbeing of different people, provided the weights are determined by the people's positions rather than by their individual identities. Since the principle of personal good seems appealing, this is an argument in favour of impartiality. Suppose, for instance, we have a value function that discounts for distance, so we give more weight to the wellbeing of our neighbours than to the wellbeing of people in India. But now suppose one of our neighbours moves to India. Suppose she is exactly as well off in India as she is at present, and suppose no one else is harmed or benefited by her move. Our value function now gives less weight to her wellbeing, so it will count her move as a bad thing. But the principle of personal good says the overall goodness of an outlook depends only on how well off the people are. Since everyone is equally as well off as before, it says that overall goodness cannot have changed. It is therefore inconsistent with our value function that discounts for distance. This creates a case against discounting.

Here is an example of a value function that is nonimpartial in a more subtle way. Inequality within a nation is commonly thought to be a worse thing than global inequality. Take a value function that reflects that view. Suppose two nations have perfect equality internally, but people in one are better off than people in the other. Now suppose one person moves from the better-off to the worse-off nation. But suppose she maintains her own standard of living in doing so. According to the principle of personal good, then, the world will be made neither better nor worse. But the person's move creates inequality in the poorer nation. What was previously a part of the inequality between nations has now become an inequality within a nation. Therefore, the world would be worse according to our inequality-averse value function. Again, the principle of personal good conflicts with this value function, because the function is not perfectly impartial.

A similar argument can be set up against a value function that discounts for time. Suppose a person might live earlier or later in time.[43] Suppose her wellbeing will be the same in either case, and the wellbeing of other people would be unaffected. Then the principle of personal good says either alternative would be equally good. But a value function that discounts for time says it is better for the person to live earlier.[44] So the principle of personal good conflicts with discounting.

Tyler Cowen has recently presented an argument like this against discounting, and a paper of Peter Hammond's refers implicitly to a similar argument.[45] Both these authors, however, rely on the Pareto principle rather than the principle of personal good. The Pareto principle is formally parallel to the principle of personal good, but it is expressed in terms of people's preferences rather than their wellbeing.[46] It says that if everyone is indifferent between two alternatives, then those alternatives are equally good.

In the examples I gave above, where a person moves from one position to another, I assumed everyone is equally well off either way. I said nothing about preferences. But Hammond and Cowen would assume that, since everyone is equally well off, everyone is also indifferent about whether the person moves or not. This would allow these authors to deduce that either alternative is equally good, using the Pareto principle. But when it comes to the intertemporal case, this way of arguing from preferences encounters a difficulty. It apparently requires a person to have a preference about when to live, before she actually does live. It is a little hard to conceive how this could happen. Certainly, there may be successful ways of getting round this difficulty, and Cowen suggests some. But it seems to me simpler to rely on the principle of personal good from the start. The principle of personal good has just as a strong a native appeal as the Pareto principle, and it does not rely on dubious preferences.

Nevertheless, I find it hard to know what weight to give the argument from the principle of personal good. If someone believes in discounting future wellbeing, she will have to give up this principle. But she may be willing to do so. She may say: 'Precisely because I value future wellbeing less than present wellbeing, I think it better to have a person living now rather than in the future, if her wellbeing will be the same either way. Therefore, I do not accept the principle of personal good.' The question is whether the principle can be given an independent defence that is strong enough to stand up to this sort of thinking. I do not know whether that is so.

It is helpful to test it out in different applications of the same style of argument. That is why I have given two examples besides the intertemporal one. But even in these more mundane examples, I have not been able to reach a conclusion about the merits of the argument. I wish to suspend judgement about it.

Discounting may protect the environment

So much for the case in favour of impartiality. To begin the case against, I shall first mention a point made by David Pearce, Edward Barbier and Anil Markandya in *Sustainable Development*. These authors mean to offer a word of warning to environmentalists who oppose discounting, rather than to produce a definite argument in favour of it. A low or zero rate of discount, they point out, encourages investment and fast economic growth. Since growth tends to damage the environment, discounting at a high rate may be a way of protecting the environment.

In practice, these authors are no doubt right. If a government uses a low rate of discount in its cost-benefit analysis, and if it maintains a low rate of interest in the economy, so that private industry discounts at a low rate in its investment decisions, then the result will be fast economic growth. This may well be bad for the environment; emissions of greenhouse gases may well increase rapidly, for instance. Furthermore, the net result may well be bad for future generations. The damage to the environment may outweigh the benefits bequeathed to them by the economic growth. That is what Pearce, Barbier and Markandya are worried about.

However, if benefits and costs are estimated using a value function like (3.1.3) or (3.1.4), it is plain that reducing the discount rate cannot possibly be bad for future generations. Reducing the discount rate simply means giving the good of future generations more weight relative to our own. So it must be beneficial to them. If reducing the rate turns out *in practice* to be bad for future generations, that can only be because of some failure in the calculations. And it is also plain what the failure is. It is to ignore environmental externalities. Private industry ignores externalities, such as the emission of greenhouse gases, and consequently, reducing the interest rate for private industry may well promote investment that harms future generations. So far as government investment is concerned, a properly conducted cost-

benefit analysis would take externalities into account.
Reducing the discount rate in a properly conducted analysis
would encourage investment only in projects that would, on
balance, benefit people living in the future. If reducing the
rate in practice harms them, that is an error of the practice.

Pearce, Barbier and Markandya are issuing a valuable
warning about the effects discounting may have in practice.
But the lesson to be drawn is that externalities must be
taken into account, not that the future should be discounted.

Risk of extinction

A good part of the case in favour of discounting is presented
in *Economic Theory and Exhaustible Resources* by P. S.
Dasgupta and G. M. Heal.[47] I shall next review these
authors' arguments.

One is that there is always a possibility of extinction.
Each generation is slightly less likely to exist that the one
before, so there is slightly less to be said for setting aside
resources for its wellbeing. In a sense, this is undoubtedly
a reason for discounting the wellbeing of future generations.
But it is no reason to have a positive discount rate r in
(3.1.3) or (3.1.4). The value function g is supposed to
represent the goodness of an outlook in which there is no
uncertainty. Uncertainty is to be accounted for separately.
It cannot affect the value of r.

No maximum

A second argument is that impartiality can run into a
particular sort of practical difficulty when faced with an
infinitely long future.[48] Imagine the earth's nonrenewable
resources are like a cake. They can be eaten up at any rate
we choose, but eventually they will be finished. After that
time, people will have to live on renewable resources, at
whatever level of wellbeing these resources permit. Suppose
that, whilst we are still depleting nonrenewables at some
rate, we shall be better off than people will be once renew-

able resources are exhausted. Suppose our wellbeing will be higher the faster we deplete, but that depletion has diminishing marginal benefits. That is to say, if we were to halve the rate of depletion, the wellbeing we derive from nonrenewables (over and above the wellbeing we would obtain from renewables only) would not be half what it was, but more than half.

How fast should we use up resources? Suppose we have an impartial value function. Then if we halve the rate of depletion, whatever it is, nonrenewables resources will last for twice as long. Twice as many people will get the benefit of them, and (because of the diminishing marginal benefits of depletion) the benefit they will get is more than half the benefit we receive at present. So total wellbeing will be greater. It follows that, whatever the rate of depletion, it is always better to halve it. And this is a problem. Certainly, we should not deplete at a zero rate, because then no one would get any benefit at all from nonrenewable resources. But on the other hand, no positive rate can be the right one, because it would always be better to halve it.

This difficulty does not arise in practice, because the earth will not exist for ever. Nevertheless, an ethical view ought to be able to stand up to counterfactual tests. But it seems to me that, if the earth were to last for ever, this argument would reveal a genuine difficulty in the nature of things, not a mere artifact of an incorrect theory of value. The theory may be correct. If it is, and if the world were to last for ever, we should have to recognize that there is no best thing to do in the circumstances. We could not maximize good. We should have to act in some other way. We might have to *satisfice* instead, for instance.[49] Instead of looking for the best possible action, we might have to be content with an action that is 'good enough'. So I do not find this a convincing argument against impartiality.

Continuity of ordering

A third argument mentioned by Dasgupta and Heal is taken from theorems established by Peter Diamond and elaborated by Tjalling Koopmans.[50] It is, once more, a problem about infinity. The beginning of it is this. If the world continues for ever, the formulae for overall good in (3.1.3) or (3.1.4) may well not be finite. Overall good may be infinite, and this is especially likely if the discount rate r is nought. Now, mere infinite sums of good need be no problem. What we need is an *ordering* of alternatives by their goodness, and that is possible even if each of the alternatives has infinite goodness overall. For instance, the alternative (2, 2, 2, . . .), which gives two units of good at every time, is better than (1, 1, 1, . . .), even though both (with zero discounting) give an infinite total.

Diamond therefore considers orderings of alternative programmes of wellbeing. Each programme maps out a sequence of wellbeings for each time to infinity. Diamond requires an ordering to be continuous, in a sense I shall shortly define. And he proves, if it is to be continuous, that it cannot be impartial: wellbeing cannot have equal weight whatever time it occurs at.[51] On the face of it, this looks like an argument against impartiality, and Dasgupta and Heal take it to be one.

But the whole weight of Diamond's proof rests on the requirement of continuity. I need to define this requirement more precisely. The condition of continuity, roughly, is this: two programmes that are close to each other should occupy nearby positions in the ordering. This condition has to be filled out by specifying when programmes are to be counted as close to each other. We need a concept of 'distance' between programmes. Diamond's definition is this: the distance between two programmes is the greatest distance by which one diverges from the other. To find the distance between two programmes, for each time calculate the difference in wellbeing between one programme and the other. The distance between the programmes is the greatest of these differences. Continuity requires that two programmes close together according to this measure are close in the ordering.

Now consider this sequence of programmes:

(1, 0, 0, 0, . . .)
(1/2, 1/2, 0, 0, . . .)
(1/3, 1/3, 1/3, 0, 0, . . .)
(1/4, 1/4, 1/4, 1/4, 0, 0, . . .)

An impartial valuation would insist that each member of this sequence is equally good, however far down the line we go. Certainly, each is better than the programme (1/2, 0, 0, 0, . . .). However, as we go down the line the members of this sequence of programmes gets closer and closer to (0, 0, 0, . . .), according to the distance measure I specified. Consequently, if the ordering is continuous, later members of the sequence will have to be close to (0, 0, 0, . . .) in the ordering. But (0, 0, 0, . . .) must be below (1/2, 0, 0, . . .). So later members of the sequence must be below (1/2, 0, 0, . . .). That is required by continuity. Impartiality, however, requires them to be above (1/2, 0, 0, . . .). Impartiality and continuity are therefore inconsistent.

In effect, that is all this argument against impartiality amounts to (though Diamond's proof actually uses a more elaborate example). Plainly, the fault lies with an unsatisfactory requirement of continuity. Diamond offers no defence of this requirement. The argument therefore raises no semblance of a valid case against impartiality.[52] A weaker continuity condition, which permits an impartial ordering, is used by Charles Harvey in 'Valuing future costs and benefits'.

Excessive sacrifice

A number of arguments against impartiality are surveyed by Derek Parfit, in order to refute them.[53] I have covered most of them already. But two are left, and I shall mention them now. The first is an argument from excessive sacrifice.

Under certain conditions, the aim of maximizing an impartial value function may impose very heavy demands on the present generation.[54] It is easy to see why this might be. Assume for the sake of argument that the saving of a life brings the same benefit at whatever time it occurs. A certain quantity of economic resources are required to save a life now, through medicine, say. If those resources were invested, either in research or simply in production where they would earn a positive return, they might well be able to save two lives, rather than one, in twenty years' time. So if our aim is to maximize an impartial value function, we ought to keep the resources to save two lives in the future, rather than use them to save one life now. We might well find that maximizing an impartial value function would lead us to abandon very many of our present life-saving activities, and instead invest the resources for the future. And this might be too much of a sacrifice to expect of the present generation.

It is a common objection to utilitarianism that it sometimes asks greater sacrifices of people than can reasonably be expected of them. This is one of John Rawls's objection, and Rawls thinks it particularly likely that utilitarianism applied between generations will demand too much of some generations.[55]

Suppose this is right. Parfit points out that it is not a reason for using a positive discount rate.[56] If excessive sacrifice should be avoided, that fact should be incorporated into the value function in a different way. A natural way would be to fix some minimum level of wellbeing below which no generation should fall. The function should attach a special large negative value to any level of wellbeing below this minimum. A function with this property would still be impartial. The argument from excessive sacrifice offers no grounds for a nonimpartial value function. Excessive sacrifice should not be asked of *any* generation, not just the present one.

Special relationships

Another common objection to impartial utilitarianism is that it does not allow people to give special value to the wellbeing of those near and dear to them. But common sense suggests we are morally permitted, and perhaps even morally obliged, to count the wellbeing our nearest and dearest for more than other people's. It also suggests we are permitted to count our own wellbeing for more than other people's.

Applied to our context, this common-sense doctrine suggests we are entitled to count our own generation for more than the next. Since the next generation consists of our children, and we can give special weight to them, we are allowed to count the next generation for more than the one that follows it. And so on: each generation will count a bit less that its predecessor. Parfit accepts that this consideration may justify some discounting. It supports a value function that is not impartial.

Nevertheless, Parfit mentions two important limits to the discounting that this common-sense idea can license. The first is this. Common sense may suggest we should give more weight to people close to us, but it does not suggest we may count other people for nothing. There is some weight we should give to any person, even the remotest stranger. But discounting at a constant rate means that the weight we give the distant future decreases towards nought. Parfit suggests this is wrong: there should be some lower limit to the weight we give it.

The second limit is that the common-sense idea does not apply to grave harms. Parfit says:

> Perhaps the U.S. Government ought in general to give priority to the welfare of its own citizens. But this does not apply to the infliction of grave harms. Suppose this Government decides to resume atmospheric nuclear tests. If it predicts that the resulting fallout would cause several deaths, should it discount the deaths of aliens? Should it therefore

> remove the tests to the Indian Ocean? I believe that, in such
> a case, the special relation makes no moral difference. We
> should take the same view about the harms that we impose
> on our remote successors.[57]

I am convinced by this argument. Releasing greenhouses
gases may impose grave harms on our successors. I do not
believe we are justified in imposing those harms simply on
the ground that we have special responsibilities to people
nearer to us.

Conclusion

That concludes my review of the arguments about the pure
discount rate. I have found fault with most of them. But I
set out from a strong initial presumption in favour of
impartiality, which is embodied in utilitarianism. Convincing arguments would be required to dislodge that presumption, and I found no convincing arguments. Only one of the
contrary arguments – the common-sense argument from
special relationships – has any validity at all, and even that
one does not justify applying a positive discount rate to
major harms imposed on future generations. I therefore
continue to favour a discount rate, applied to pure harms
and benefits, of nought.

Notes

1. There is an introduction to separability conditions in my *Weighing Goods*,
Chapter 4.
2. Exponential discounting can be derived from an axiom of Koopmans's
called 'stationarity' ('Representation of preference orderings over time'), but
I see no particular appeal in the axiom. It is also true that a nonconstant
rate may lead a government to change its mind as time progresses (see
Strotz, 'Myopia and inconsistency in dynamic utility maximization'). This
suggests there may be something wrong with a nonconstant rate, but I do
not know if this point can be developed into a proper argument in favour
of exponential discounting.
3. Cowen and Parfit draw attention to the importance of separating these

questions.

4. See pp. 116–121. Total and average utilitarianism are both forms of (3.1.4).

5. This view is defended by McKerlie in 'Equality and time' and by Temkin in 'Intergenerational inequality'.

6. 'The rate of discount for benefit-cost analysis.'

7. See p. 20 above.

8. An exception is the discussion of Diamond and Mirrlees on pages 83–90.

9. This is oversimplified. For it to be definitely true, the changes must be small, and the consumer must not be at a kink or a corner of her indifference curve.

10. See, for instance: Dasgupta, 'Resource depletion, research and development, and the social rate of discount', pp. 277-9; Pearce, 'Ethics, irreversibility, future generations and the social rate of discount', pp. 80–4; Pearce, Barbier and Markandya, *Sustainable Development*, p. 50.

11. See, for instance, Graaf, *Theoretical Welfare Economics*, pp. 90–2.

12. Compensation between generations (see, for instance, Pearce, 'Ethics, irreversibility, future generations and the social rate of discount') is not at issue here. If a project is profitable at consumer prices, that means compensation is possible between present consumers to ensure that none of them is harmed. But future generations do not enter the story at all, a fact I shall be objecting to on pp. 71–74.

13. *Wealth of Nations*, Book 1, Chapter 5 (Everyman edition, p. 28).

14. When the UK Department of Transport, for instance, does a cost-benefit analysis for a road, it includes a figure for deaths amongst 'accident costs', and discounts it along with other costs. (See its *Values for Journey Time Savings and Accident Prevention*.) The US Office of Management and Budget (interpreting Executive Order 12291) requires US agencies to use a similar procedure. But at least one member of the UK Treasury is aware of the mistake; see Spackman, 'Discount rates and rates of return in the public sector', p. 11.

15. 'The social rate of discount and the optimal rate of investment', p. 97.

16. *The Economics of Welfare*, pp. 29–30.

17. 'The social time preference discount rate in cost-benefit analysis', p. 251 in the reprinted version.

18. Arrow, *Social Choice and Individual Values*. See also my *Weighing Goods*, Chapter 7.

19. Marshall, *Principles of Economics*, p. 120; Pigou, *The Economics of Welfare*, p. 25. Notice that to say people are imprudent is not necessarily to say they are irrational. Pigou thought imprudence was irrational, but Marshall did not. Derek Parfit has recently argued that imprudence may be rational. (*Reasons and Persons*, Part II.)

20. See Lind, 'A primer on the major issues relating to the discount rate', pp. 83–4.

21. Marshall was aware of this point (*Principles of Economics*, p. 121), but

decided to ignore it. In 'Positive time preference', Olson and Bailey claim to offer compelling evidence that people are imprudent. Much of their case depends on mistaken mathematics: readers will notice that on their p. 12 these authors forget that the marginal benefit of consumption can decline towards nought in the limit without actually ever becoming nought. Setting that error aside, their evidence amounts to this. On p. 66 above, I explained that (if there is no uncertainty) the consumer interest rate exceeds a consumer's pure rate of discount by an amount equal to the rate at which the consumer's marginal benefit of consumption is declining. On the basis of some very sketchy figures (their p. 19), Olson and Bailey suggest it is implausible that people's marginal benefit of consumption is falling at a rate as high as the consumer rate of interest. Consequently, they deduce, people's pure rate of discount must be positive. This scarcely amounts to 'compelling' evidence.

22. Marglin, 'The social rate of discount and the optimal rate of investment'. Sen has written very many papers on the subject, of which I list only these two: 'Approaches to the choice of discount rates for social benefit-cost analysis', and 'Isolation, assurance, and the social rate of discount'.

23. p. 83.

24. Note that what I am calling the 'direct' inputs here are all inputs and outputs apart from the influence on private investment. Ordinary externalities, such as emissions of carbon dioxide, count as direct inputs or outputs. This significance of this point, if it is not obvious, is emphasized on page 101 below.

25. The literature is very nicely summarized by Lind in 'A primer on the major issues relating to the discount rate'.

26. 'The rate of discount for benefit-cost analysis.'

27. 'Reassessing the government's discount rate policy'.

28. p. S13.

29. 'On the private provision of public goods'.

30. 'The isolation paradox and the discount rate for benefit-cost analysis'.

31. Sen, Introduction to *Resources, Values and Development*; Newbery, 'The isolation paradox'.

32. See 'Optimal taxation and public production', pp. 18–19.

33. *The Methods of Ethics*, p. 414.

34. This view is implicit in most of the recent debate about 'consequentialism'. It is almost explicit in Parfit, *Reasons and Persons*, p. 27. There is an explicit defence of it in Regan 'Against evaluator relativity'.

35. Many authors fail to distinguish between impartiality and agent neutrality, and seem to assume that good must be impartial. Scheffler, for instance, suggests that good should be 'judged from an impersonal standpoint which gives equal weight to the interests of everyone'. (Introduction to *Consequentialism and Its Critics*, p. 1.)

36. See p. 42.

37. *Rational Behavior and Bargaining Equilibrium*, pp. 48–9.

38. 'A mathematical theory of saving', p. 261 in the reprinted version.
39. *Reasons and Persons*, pp. 480–6.
40. 'The economics of resources or the resources of economics', p. 9.
41. p. 9.
42. *A Theory of Justice*, p. 294.
43. Now that viable embryos can be stored, this is a practical as well as a conceptual possibility.
44. Actually, this is not necessarily so. It depends on other aspects of the value function as well as discounting. The function might be nonimpartial in a second way that cancels out the effects of the first. Suppose, for instance, that the function is sensitive to inequality within a cohort. If the person lives earlier, that might create more inequality within a cohort than if she lives later, and this might cancel out the effect of the discounting. The value function expressed in (4.1.2), which values average wellbeing within cohorts, rather than total wellbeing, might also prefer a person to live later rather than earlier, even if it discounts the average wellbeing of later cohorts. However, it would be easy to reconstruct the argument against discounting, using a more complicated example.
45. Cowen, 'A positive argument for a zero rate of intergenerational discount'; Hammond, 'Consequentialist demographic norms and parenting rights', pp. 137–8. Cowen's argument is more elaborate than mine, and buttressed against various objections.
46. See my *Weighing Goods*, Chapter 7.
47. pp. 255–82.
48. My simple example is taken from *Economic Theory and Exhaustible Resources*, pp. 267–8. There is a more elaborate example in Dasgupta 'Resource depletion, research and development, and the social rate of discount'.
49. Ethical satisficing has been recommended in much less awkward circumstances than this by Slote in *Beyond Optimizing*. Although it is not maximizing, satisficing should be included within teleology. See my *Weighing Goods*, p. 7.
50. Diamond, 'The evaluation of infinite utility streams'; Koopmans, 'Representation of preference orderings over time'. Page, as well as Dasgupta and Heal, gives some weight to this argument, in 'Intergenerational equity and the social rate of discount' and 'Intergenerational justice as opportunity'.
51. Actually, Diamond attributes one of his main theorems to M. E. Yaari.
52. Diamond also considers an alternative definition of continuity. But the alternative is even less satisfactory, because it incorporates a positive rate of discount even into the measure of distance. Koopmans (p. 92) mentions another way to avoid Diamond's objection to impartiality: to ask for only a partial ordering of the programmes, rather than a full ordering. This might, indeed, be a satisfactory move. But it is not needed, because Diamond's objection is anyway empty.

53. *Reasons and Persons*, pp. 480–6. The same points appear in Cowen and Parfit, 'Against the social discount rate'.

54. See Dasgupta and Heal, *Economic Theory and Exhaustible Resources*, p. 261, and the references mentioned there.

55. *A Theory of Justice*, p. 287.

56. Rawls agrees; see *A Theory of Justice*, p. 297.

57. *Reasons and Persons*, p. 486.

Chapter 4
Aggregating Wellbeing

The problem I posed in Chapter 2 is to discover the form of the value function that aggregates wellbeing across a distribution. In principle, we need to deal with three-dimensional distributions, where the dimensions are people, time and states of nature. But I intend to concentrate on people and time only. We need to bear in mind that, in the time dimension, the length of a person's life may vary and, in the dimension of people, the population of the world may vary. Global warming will cause major variations of this sort. What is needed is an integrated treatment of aggregation across the two dimensions, which takes these demographic variations into account. Indeed, we really need an integrated treatment of aggregation in three dimensions, including the dimension of states of nature.[1]

Chapter 3 examined one aspect of the problem: the question of discounting. This is a question about the aggregation of wellbeing across the dimension of time. But the literature on discounting treats only this one aspect, without taking account of the other features of the problem. This leads it into error. It typically assumes that wellbeing is strongly separable across time. And I explained in Section 3.1 that, when one takes account of variations in the length of people's lives, this is an implausible assumption.

There has been a lot of discussion of other aspects of the problem taken individually. There is, for instance, a large literature on 'the value of life' – on the benefit of extending a person's life, that is.[2] There is also a large literature on the value of changes in population.[3] Another important aspect of the problem is the value of equality – a matter of how different people's wellbeing should be aggregated together – and this too has been much discussed. I know of a certain amount of work that takes together more than one aspect of the problem,[4] but none that treats the problem comprehensively. I am not able to provide a comprehensive treatment myself; this, I believe, should be the main focus for future research. Furthermore, the work that has been done on the value of population changes demonstrates clearly that satisfactory conclusions are hard to come by. Therefore, what I can say in this chapter will inevitably be rather disconnected and inconclusive. I shall review various conditions that might be imposed on the value function, and ask how plausible they are.

Since I have found little justification for discounting wellbeing by time, I shall not include discount factors in this chapter. Anyone who wants to add them can do so, without making any difference to the main argument. I shall also not concern myself with the value of equality. Again, any one who wants to incorporate a value for equality into the value function may do so easily.[5] I shall concentrate particularly on the demographic aspects of the value function: the aspects that particularly concern the size of the population and the lengths of people's lives.

One reason for concentrating on these demographic aspects is that global warming will cause demographic changes. But that is not the only reason. Even if global warming were not to influence demography at all, demographic variables will nevertheless certainly change a great deal over the period of time we are concerned with. The world's population is growing, and life expectancies are generally increasing. Consequently, the way the demographic variables enter the value function can make a great difference to the value the function will assign to alternative outlooks. Even changes that have no demographic effects themselves will be valued differently according to how demography enters the value function. On page 119, for instance, we shall see that total utilitarianism will have different implications from several versions of average utilitarianism.

The conditions that may be imposed on the value function fall into two classes: high-level and low-level. High-level conditions fully determine the value function; low-level conditions only some particular feature of it. Section 4.1 considers various high-level conditions, and Section 4.2 some low-level conditions.

4.1 Types of utilitarianism

The high-level conditions I shall consider all happen to be versions of utilitarianism.

Complete utilitarianism

The first I call *complete* utilitarianism. Complete utilitarianism claims that the value function has the form given in (2.3.4) on page 49, which I reproduce here:

$$g = \sum\nolimits_{(i,\, t \,\mid\, g_t^i \,\neq\, \Omega)} g_t^i$$

This formula says that the value of an outlook is simply the total of everybody's wellbeing at every time. Clearly, complete utilitarianism implies weak separability between times, which is expressed in (3.1.1). I have already raised an objection to complete utilitarianism, therefore: I said on page 54 that weak separability between times is not very plausible. I can now put the reason more generally. Complete utilitarianism values the total of wellbeing in the world. There are two ways the total can be increased. One is by extending or improving the life of an existing person. The other is by bringing more people into existence. According to complete utilitarianism, either method is equally good. But intuition suggests this is wrong. Intuition suggests it is better to bring benefits to someone who exists, rather than to bring someone into existence who would have the same benefits. Compare alternatives *A* and *B* in Figure 3 on page 55. The choice between them is a matter of whether to preserve the life of one person, or to allow that person to die and have another person come into existence instead. Intuition suggests that the former is better. But according to complete utilitarianism, these alternatives are equally good.

Total utilitarianism

An alternative that does not necessarily have this implication is *total* utilitarianism. Total utilitarianism says that the value of a distribution is the total of people's wellbeing:[6]

$$(4.1.1) \qquad g = \sum_{\{i \mid g^i \neq \Omega\}} g^i$$

where g^i is the lifetime wellbeing of the ith possible person (which I take to be Ω if this possible person actually never lives). This value function will coincide with complete utilitarianism if each person's wellbeing is assumed to be simply the sum of her wellbeing over time. But we need not assume that. If we allow a person's lifetime wellbeing to include an element for the length of her life, as well as the total of wellbeing she enjoys during her life, that will overcome the objection I have mentioned. It will allow us to value *A* above *B* in Figure 3, and also *C* above *D*.

Nevertheless, total utilitarianism is commonly thought to be unacceptable too. The common objection to it is what Derek Parfit calls 'the repugnant conclusion'.[7] Imagine a world in which everybody leads lives that are very good. Compare it with a world possessing a much larger population, but where each person's life is not very good at all; it is just worth living, but only just. Provided this second world has a big enough population, it will have a greater total of people's wellbeing than the first world. So according to total utilitarianism, it is better. But Parfit, for one, finds this conclusion repugnant.

Average utilitarianism

Thoughts like this have led many authors to abandon total utilitarianism in favour of some alternative. In economics, the alternative most commonly adopted is some sort of *average* utilitarianism. Instead of taking the total of people's wellbeing as an objective, economists very often take average wellbeing. But the averaging can be done in many different ways, and each gives a different value function.

One way is to take an average of everybody's wellbeing at a particular time, and then add up the averages across time. The value function is:

$$g = \sum_t \frac{1}{n_t} \sum_{\{i \,|\, g_t^i \neq \Omega\}} g_t^i$$

where n_t is the number of people living at t. This is the version of average utilitarianism most commonly found in economics.[8] But this function is weakly (indeed strongly) separable over time. It is therefore subject to the very same objection as I raised to complete utilitarianism. It moves us no further forward. Indeed it moves us very far backwards. If someone's level of living is below the average level at the time, this formula makes the death of this person count as an improvement. It would be in favour of floods that kill impoverished Bangladeshis, because their lives are probably less good than the average life of people in the world. This is absurd. So this form of the value function must be rejected.

To avoid such an absurd conclusion, the averaging will need to be done taking whole lives together. It could, for instance, be done one cohort at a time. For each cohort, we could take the average lifetime wellbeing of the people who make it up, and take our value function to be the total, over time, of the averages for each cohort. That is:

$$(4.1.2) \qquad\qquad g = \sum_t \frac{1}{N_t} \sum_{i \in C_t} g^i$$

where C_t is the tth cohort and N_t the number of people in it.

This principle, however, is also unsatisfactory. Compare the following two policies. Under either, the same number of people live altogether. And each cohort has the same average wellbeing under one as it has under the other. But the average changes over time, and the sizes of the cohorts are not the same under the two policies. Under one, the cohorts are larger at times when wellbeing is high; under the other they are larger when wellbeing is low. That is to say, one policy brings it about that more people live when living standards are high; the other when they are low. According to (4.1.2), these two policies are equally good. But it seems clear that they are not equally good. It seems clearly better that more people should enjoy higher living standards, and that will occur under the first policy.

The trouble with (4.1.2) is that, because it averages cohorts separately, it gives more weight to the wellbeing of someone who lives in a small cohort than it does to the wellbeing of someone who lives in a large one. One consequence of this defect is that (4.1.2) is not an *impartial* value function. More specifically, it is not impartial between times.

Figure 6

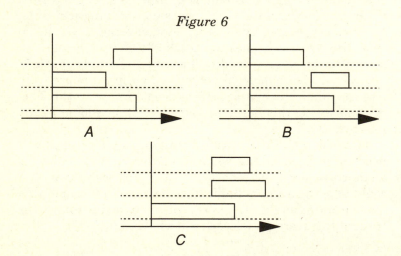

I have already considered one notion of temporal impartiality in Section 3.4. I now want to introduce a slightly different notion. I shall say that a value function is temporally impartial if and only if its value is unaffected by transferring a life forwards or backwards in time. Consider Figure 6. Temporal impartiality requires that alternatives A and C are equally good, because the only difference between them is that one person's life is displaced in time. Equation (4.1.2), on the other hand, says C is better. It is hard to believe that C is really better. Someone who believes in discounting future benefits, and therefore rejects impartiality, would probably think that C is actually worse than A, because it delays benefits.

Since the world's population is growing, (4.1.2) values a benefit brought to a later person for less than the same benefit brought to an earlier one. Compared with the impartial total utilitarian formula (4.1.1), it would be less inclined to favour long-term projects. In this way, this value function differs from total utilitarianism even over projects that themselves have no demographic effects. This is an example of an important point I made on page 114. The way demography enters into the value function makes a difference, not just to how the function values demographic changes, but to how it values any long-term changes at all.

The only way to overcome the lack of impartiality of (4.1.2), within the spirit of average utilitarianism, is to make the averaging cross time; we shall have to average one cohort with another. But it is very hard to know how far across time the averaging should extend.[9] Suppose we make our value function the average of everyone's wellbeing:

$$g = \frac{1}{N} \sum_{\{i \ | \ g^i \neq \Omega\}} g^i$$

where N is the total number of people. Then it is hard to know who should be included in the number N. This makes a significant difference. According to this value function, it is a good idea to add more people to the world's population if and only if those people are better off than the existing average. Life in the stone age, let us assume, was hard. So if we include stone-age people in the average, the average will be low, and we shall be more inclined to favour increases in the world's population. But it seems absurd that conditions of life in the stone age should influence the value we attach to our present policies. So perhaps we should only include present and future people in the average. But then the value of the policy will depend on what date it is evaluated at, since that determines which people will be counted. This seems a peculiar consequence. As Peter Hammond points out,[10] the result is that average utilitarianism is dynamically inconsistent.

The fundamental difficulty is simply that it is very odd to make the value of a person – that is to say, the value of having a particular person come into existence – depend on the wellbeing of other people, who may be remote from her. More technically: no average utilitarian formula is strongly separable between possible people, as I shall be saying on page 127. This seems to me sufficient grounds for rejecting average utilitarianism. There is very little basis for averaging principles anyway. Few arguments have been offered in their defence. Averaging principles seem to have come into existence amongst economists as an ad hoc device for avoiding the repugnant conclusion.[11]

It is true that John Rawls prefers average utilitarianism to total utilitarianism.[12] His argument is this. Suppose people are placed in an 'original position', where they have to choose the type of society they would like to live in. Suppose they know they will live in whichever society they choose, but they do not know which position in the society they will occupy. Their position will be determined randomly, and they have an equal chance of being anywhere. Then, if they are expected utility maximizers, they will choose a society where the average level of wellbeing is highest. Rawls believes this gives some support to average utilitarianism. However, Brian Barry and others have convincingly demonstrated that an original position argument like this cannot properly be applied when the size of the population is in question.[13] I therefore attach no weight to this argument.

One other point bears mentioning. Total and average utilitarianism are very different theories, and where they differ most is over extinction. If global warming extinguishes humanity, according to total utilitarianism, that would be an inconceivably bad disaster. The loss would be all the future wellbeing of all the people who would otherwise have lived. On the other hand, according to at least some versions of average utilitarianism, extinction might not be a very bad thing at all; it might not much affect the average wellbeing of the people who do live. So the difference between these theories makes a vast difference to the attitude we should take to global warming. According to total utilitarianism, although the chance of extinction is slight, the harm extinction would do is so enormous that it may well be the dominant consideration when we think about global warming. According to average utilitarianism, the chance of extinction may well be negligible.

Other versions of utilitarianism

Two more recent versions of utilitarianism were originally offered as ways of avoiding the repugnant conclusion, without taking on the unacceptable implications of average utilitarianism. One is 'critical level utilitarianism', whose value function is

$$(4.1.3) \qquad g = \sum_{\{i \mid g^i \neq \Omega\}} (g^i - \alpha)$$

The number α in this formula is the 'critical level'. The formula says the value of a distribution is the total over all people of the amounts by which each person's wellbeing exceeds the critical level. Suppose a person might be added to the population. If her wellbeing would be above the critical level, then (4.1.3) is in favour of adding her; if it would be below, then (4.1.3) is in favour of preventing her existence. This theory is recommended by Charles Blackorby and David Donaldson in 'Social criteria for evaluating population change'.

It is what Parfit calls 'an appeal to the valueless level', and Parfit offers an objection to it.[14] His objection is this. Instead of the repugnant conclusion, critical level utilitarianism leads to the following parallel conclusion. Imagine a world where people live very good lives. There is always another world, containing very large numbers of people living lives just above the critical level, which the theory says is better. If the critical level is low, close to the level at which life is only just worth living, then this conclusion is not much different from the repugnant conclusion, and not much more plausible. But suppose, on the other hand, that the critical level is higher. Adding people to the world whose lives are below this level is, according to the formula, a bad thing. But these people's lives are not bad ones. They are worth living, even though they are below the critical level. Yet critical level utilitarianism would be positively opposed to these people's existence. It would recommend definite sacrifices on the part of existing people, making them worse off, for the sake of preventing these people from living. And that would be very strange.[15] The theory is caught in a dilemma, then. Set the critical level too low and the theory is implausible for one reason; set the level too high and it is implausible for another.

A second new version of utilitarianism is Yew-Kwang Ng's 'number-damped utilitarianism'.[16] Its value function is:

$$g = \frac{\phi(N)}{N} \sum_{\{i \mid g^i \neq \Omega\}} g^i$$

where ϕ is an increasing concave function. This formula is like total utilitarianism, except that the value of adding people to the population diminishes as the population increases. But this formula has the same basic failing as average utilitarianism. It is not strongly separable between possible people. It makes the value of a person depend on how many other people there are, and how well off they are, even if those other people are remote in time and space. And this is implausible.[17]

4.2 Other conditions

Now I turn to low-level conditions on the value function.
These are conditions that do not purport to determine the
function completely. It will soon become apparent, however,
that combinations of them may well go a long way towards
determining the form of this function. A natural way for
research in this area to proceed would be to look for conditions that can be supported by good arguments, and consider
what implications they together have for the value function.
We may hope to end up with a set of defensible conditions
that together determine the form of the value function
completely. This – the traditional approach of social choice
theory – has been applied to population theory by Charles
Blackorby and David Donaldson, and by Peter Hammond.[18]
Much of my argument below follows the lead of these
authors.

Before I come to that, though, I need to mention a
condition that is appealing but has in the end to be rejected.

The person-affecting condition

I mentioned that average utilitarianism is one response to
the repugnant conclusion. I rejected it, largely on the
grounds that it has no particular principle behind it. An
alternative response originates in a paper of Jan Narveson's
and is, on the face of it, more attractive.[19] It starts from an
intuitive diagnosis of what is wrong with total utilitarianism. Total utilitarianism cares about wellbeing, but it does
not care about who has that wellbeing. It elevates wellbeing
to an object in itself. Consequently, if it turns out that the
best way of increasing wellbeing is to have a lot of people,
then that is what total utilitarianism favours. But the
intuition that attracts us to utilitarianism in the first place
is that we believe people should have good lives. We think
people should be well off, but we do not value wellbeing for
its own sake. 'We are in favour of making people happy, but
neutral about making happy people,' Narveson says.[20]

Suppose, for instance, a couple are thinking about having a child. Then the fact that the child would be happy if they had it is not in itself a reason for having one, and they would be doing nothing wrong if they decided not to. Also, we would not be in favour of sacrificing the wellbeing of existing people for the sake of adding new people to the world.

All this is intuitively appealing. How might it be formulated as a condition on the value function? Suppose we have to compare two alternative actions. Suppose one contains a certain number of people, and the other contains all the same people, and some more as well. This, for instance, is the choice facing a couple wondering whether to have a child. Then the condition is that one alternative is at least as good

Figure 7

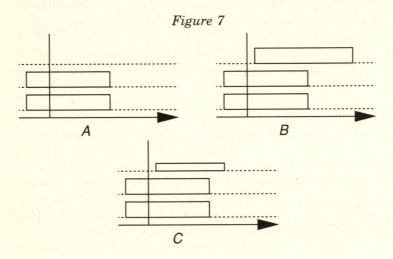

as the other if and only if it is at least as good for the people who exist in both. Figure 7 shows a simple example. The condition says that alternatives *A* and *B* are equally good, because they are equally good for the two people who exist in both. The wellbeing of the person who exists only in *B* does not affect the comparison. Using Parfit's terminology, I call this 'the person-affecting condition'.[21]

Unfortunately, however, the person-affecting condition is self-contradictory. In Figure 7, it says, not only that A and B are equally good, but also that A and C are equally good, since A and C are equally good for the two people who exist in both. The relation 'equally as good as' must be transitive. Therefore, B and C are equally good. But the condition also says that B is better than C, since it is better for one of the three people who exist in both B and C, and worse for none of them. So here is a contradiction. Various attempts have been made to rescue the condition from this difficulty, but I do not believe any have been successful.[22] The person-affecting condition seems doomed to fail.

This is not an easy conclusion to come to terms with. The intuition underlying the person-affecting condition is a very strong one. It is not at all easy to accept that the wellbeing of existing people should sometimes be sacrificed for the sake of adding new people to the world, or that a couple are doing anything wrong if they decide not to have a child who would be happy. In economics, almost the entire body of literature on valuing human life rests on the assumption that adding people to the population has no value in itself: if a person's life is saved, the extra lifetime of the person herself is valued, but not the lives of any children she later has.[23] If the person-affecting condition goes, all of this has to go with it.

Separability

What other low-level conditions might be imposed?

I myself accept, first, the principle of personal good, which is expressed in (2.3.3) on page 47. Equation (2.3.3) shows that this principle is a condition of weak separability between the rows in a diagram of the wellbeing distribution such as Figure 2 on page 46. The value function is weakly separable between possible people, that is. For reasons I gave on page 47, I feel secure in this condition.

I also favour a second separability condition. In principle this one ought to be outside the scope of this study, but it needs to be included for a reason that will soon appear. On page 48, I explained that I would concentrate on two-dimensional distributions of wellbeing, and ignore the dimension of states of nature. I have left uncertainty out of this study. However, I do now need to mention one point about the third dimension. I accept expected utility theory. And expected utility theory implies a condition of separability known as the 'sure-thing principle' or 'strong independence axiom'.[24] This condition says that the value function is strongly separable by states of nature. It is therefore weakly separable by states of nature. This weak separability is the condition I need now.

The value function, then, is weakly separable both by possible people and by states of nature. The combination of these two separability conditions is significant. It allows us to apply a theorem of W. M. Gorman's.[25] This theorem tells us that if the value function is weakly separable between possible people, and also weakly separable between states of nature, then it is *strongly* separable between possible people.[26] Strong separability is equivalent to additive separability.[27] Therefore, the value function has the form:

$$(4.2.1) \qquad g = \sum_i f^i(g_1^i, g_2^i, g_3^i, \dots)$$

This equation rules out any sort of average utilitarianism. In that, it accords well with intuition. In effect, strong separability says that the value we assign to what happens to one person is independent of what happens to anyone else. For one thing, the value we assign to one person's existence is independent of how well off anyone else is. But average utilitarianism makes the value of one person's existence depend on the effect this person's existence has on average wellbeing. And this depends on the wellbeing of other people. So average utilitarianism is inconsistent with strong separability. This was the feature of it I objected to on page 120. For the same reason, (4.2.1) rules out number-damped utilitarianism, and again that accords with intuition.

Impartiality

Equation (4.2.1) does much more than rule out average utilitarianism. To display its full implications, let us add a particular condition of impartiality to the two separability conditions that have brought us this far. Let us add the condition that the value of a distribution is unaltered if the rows of the distribution are permuted. This condition says it does not matter which possible people come into existence, and also, if a life of a particular quality is lived at a particular time, it does not matter who lives it. Identity does not matter; only the quality and date of lives. This is not the same as the temporal impartiality condition I defined on page 119. It does not insist that the value of a distribution is unaffected by changing the temporal position of a person. In Figure 6, it insists that A and B are equally good, but not that A and C are. This seems to me a very plausible condition. The only reason I know for doubting it is the common sense view I mentioned on page 107. Common sense suggests we are allowed, or even that we are obliged, to count the wellbeing of people near to us for more than we count the good of strangers. I accept there may be something in this view. But I gave reasons for setting it aside in the case of intertemporal impartiality, and now, at least for the sake of argument, I shall set it aside here too.[28]

This impartiality condition will ensure that, in (4.2.1), the functions f^i are the same for every i. For a possible person who never exists, let us set f^i to be nought. That is to say, $f^i(\Omega, \Omega, \ldots) = 0$. This is simply an arbitrary matter of normalization; it makes no difference to (4.2.1) which level we choose for $f^i(\Omega, \Omega, \ldots)$.

The next thing to notice is that, if i is a person who lives at some time, then f^i must be a function of i's wellbeing g^i. This follows from combining (4.2.1) with the principle of personal good (2.3.3). So

(4.2.2) $$g = \sum_{\{i \mid g^i \neq \Omega\}} f(g^i)$$

An argument that stems originally from John Harsanyi[29]

suggests f will have to be a linear function. This argument
is open to doubt. One might be inclined to make f a strictly
concave function, since that is a way of giving some value to
equality in the distribution of wellbeing between people. But
I am not particularly concerned with equality in this
chapter. I shall therefore refrain from reviewing the argument, and simply take it for granted that f must be linear.
That is to say, $f(g^i)$ is an increasing linear transform
$\beta(g^i - \alpha)$ of g^i. The scaling factor β of the transformation
makes no difference to the sum in (4.2.2), and we can
normalize it to one. We are left with $f(g^i) = g^i - \alpha$. This
makes (4.2.2):

$$g = \sum_{\{i \,\mid\, g^i \,\neq\, \Omega\}} (g^i - \alpha)$$

which is Blackorby and Donaldson's critical level utilitarian
formula (4.1.3).

Utilitarianism again

So our combination of low-level conditions has led us back
to a version of utilitarianism. The formula we have arrived
at is subject to the objection I raised on page 122. On the
other hand, I find it hard to doubt the two separability
conditions I have mentioned, nor the impartiality condition,
and these conditions together lead us almost inexorably to
(4.1.3). Plausible conditions lead to an implausible result. In
the area of population theory this happens often. Conditions
that seem individually very plausible very easily get into
conflict with each other. I am not happy with (4.1.3), but I
find myself forced towards it.[30] I believe there is room for
much more research in this area.

Notes

1. See page 48.
2. For instance, Jones-Lee, *The Economics of Safety and Physical Risk*.
3. The papers cited later in this chapter constitute a small sample.
4. For instance, Arthur, 'The economics of risk to life', and my 'The economic value of life' both treat the value of life and the value of population changes together.
5. Blackorby and Donaldson show how in 'Social criteria for evaluating population change'.
6. Total utilitarianism is associated with Sidgwick; see *The Methods of Ethics*, pp. 414–6.
7. *Reasons and Persons*, p. 388.
8. For instance, Eckstein, 'Investment criteria for economic development'. See the discussion in Lecomber, *The Economics of Natural Resources*, pp. 69–71.
9. This problem is described by McMahan in 'Problems of population theory', and also by Hammond in 'Consequentialist demographic norms and parenting rights'.
10. 'Consequentialist demographic norms and parenting rights', pp. 133–4.
11. Sidgwick attributes them to 'political economists of the school of Malthus' (*The Methods of Ethics*, p. 415). There is a history in Sumner, 'Classical utilitarianism and the population optimum', p. 107.
12. *A Theory of Justice*, pp. 161–7.
13. Barry, 'Rawls on average and total utility: a comment'.
14. *Reasons and Persons*, pp. 412–6.
15. There is a similar objection in Ng, 'Social criteria for evaluating population change'.
16. 'Social criteria for evaluating population change'.
17. This objection is raised by Cowen in 'Normative population theory'.
18. Blackorby and Donaldson, 'Social criteria for evaluating population change'; Hammond, 'Consequentialist demographic norms and parenting rights'.
19. 'Utilitarianism and new generations'.
20. 'Moral problems of population', p. 73 in reprinted version.
21. *Reasons and Persons*, p. 394.
22. In my paper 'Some principles of population', there is an attempt to rescue it by means of a notion of 'conditional betterness'. In effect, this notion is intended to allow for intransitivity in the relation 'equally as good as'. But an intransitivity in the relation 'better than' is intolerable, and it is easy to extend the example to create an intransitivity of that sort. 'Some principles of population' does so. In 'Intransitivity and the mere addition paradox', Temkin uses a similar example, derived from Parfit's 'mere addition paradox' (*Reasons and Persons*, pp. 419–441) to argue that the relation 'better than' is actually intransitive. But I believe this is a logical

contradiction (see *Weighing Goods*, pp. 11–12). In 'Lives and well-being' and 'Population size and the quality of life', Dasgupta accepts a modified person-affecting condition, and describes a procedure for decision making that always arrives at a definite decision even in the face of examples like the one I have given. But Dasgupta fails to resolve the apparent intransitivity in the relation 'better than'. His procedure will often lead to the choice of an alternative that, according to the person-affecting condition, is worse than another of the alternatives available. So this is no resolution of the problem. (See the discussion in 'Some principles of population'.) In a very recent paper, 'Population and savings', written after this report, Dasgupta modifies the person-affecting condition still further. By doing so, he avoids intransitivity. I believe, however, that his new condition is so much modified that it has lost the intuitive appeal that makes the person-affecting condition attractive in the first place. I hope to give this recent paper a proper discussion elsewhere, but unfortunately I cannot do so here.

23. See my 'The economic value of life'.

24. See, for instance, Savage, *The Foundations of Statistics*, pp. 21–6.

25. 'The structure of utility functions'.

26. In 'Social criteria for evaluating population change', Blackorby and Donaldson take an alternative route to arrive at the same conclusion that the value function is strongly separable between people. They do not rely on expected utility theory. But it seems to me that, in effect, they beg the question. In effect, they assume strong separability from the start.

27. See the account of separability in my *Weighing Goods*, Chapter 4.

28. There is another reason that might lead one to doubt this impartiality condition. But it is one that arises from the person-affecting condition, which I have already rejected. I therefore do not take it seriously. There is a discussion of it in Parfit, *Reasons and Persons*, p. 357–78.

29. 'Cardinal welfare, individualistic ethics, and interpersonal comparisons of utility'. See the discussion in my *Weighing Goods*, Chapter 10.

30. One final point is worth mentioning. Suppose we were to assume that the value function is weakly separable by times. This assumption, together with the principle of personal good and the impartiality condition would take us, through Gorman's theorem, to complete utilitarianism. But I think we can safely reject complete utilitarianism, because I see no appeal in the assumption that the value function is weakly separable by times.

Chapter 5
Conclusions

Global warming raises many difficult questions about our responsibility to the future. In this study, I have tried to bring some order to the questions, rather than find answers to them. I have tried to map out the territory that needs to be explored, review the exploration that has already been done, and suggest some directions for future research. This chapter is a summary of my principal conclusions.

In Chapter 1, I outlined the present state of scientific opinion. The changes brought about by global warming are likely to be very large, and they will take place over a very long period of time. They are also very uncertain: the changes in climate are hard to predict, and the effects on human life much more so. These features put the problem of global warming beyond the normal experience of economics. We cannot expect the established methods of economics to handle them adequately. Consequently, research cannot be limited simply to applying, say, the standard techniques of cost-benefit analysis to global warming. In this new context, for instance, we cannot rely on any of the conventional methods for fixing a discount rate. Fundamental theoretical work is needed first. This is not to say that action in response to global warming can be delayed. Governments need to act, but they must not think that their decisions can be made for them by straightforward techniques such as cost-benefit analysis.

In Chapter 2, I described the general form of the question I believe needs to be answered, if the problem of global warming is to be tackled properly. Different actions we might take will distribute wellbeing differently across generations and across the people in each generation. Somehow, the wellbeing of all the individual people in all generations together determines the overall value of the distribution. The question is: how? How is people's wellbeing aggregated to determine the overall value of a distribution? How, specifically, is one generation's wellbeing to be weighed against another's? We need, I said, a 'value function', which determines the overall value of a distribution, on the basis of each person's wellbeing at each time.

The question I posed, then, is about wellbeing rather than justice. It is not about the rights of future generations, but about their good. It is, as I put it, a teleological question. I explained in Chapter 2 that much recent thinking about justice does not give a persuasive account of our responsibilities to future generations. That is why I think we ought to concentrate on wellbeing.

Also in Chapter 2, I mentioned one feature of the distribution of wellbeing that we need to be particularly concerned with. Global warming, and actions to mitigate global warming, will not simply alter the wellbeing of those people who live, while they live. They will certainly affect the lengths of many people's lives, and they will influence the population of the world: how many people will live in the future, and which people they will be. In comparing alternative actions, we shall need to compare outlooks with different numbers of people living for different lengths of time. Our value function will need to take account of 'demographic changes' as I called them: changes in population and the length of lives.

In Chapter 3, I examined one much discussed aspect of aggregation across time: discounting the future. The discussion of discounting has sometimes been a little confused because discount factors may be applied to different things. One may ask whether future *wellbeing* should be discounted, so that the wellbeing of future generations should count for less than our present wellbeing. Economists, though, are most often looking for discount factors that can be applied to future *commodities*, either commodities consumed or commodities produced. Suitable discount rates for commodities can sometimes be found from the market, either from the consumer interest rate or the producer interest rate. Chapter 3 considered when and how this can be done. When the method works, it can spare us the trouble of fixing on a discount rate for wellbeing; it offers us a short cut past this problem. But I argued that, in the context of global warming, the method will not work. Against the consumer interest rate, I pointed out that future people do not participate in the market. Consequently, the consumer rate does not properly reflect the value of future commodities. Against the producer rate, I pointed out that the production of commodities releases greenhouse gases and damages the environment in other ways. It therefore has negative external effects, which are not registered in the producer interest rate. Consequently, this rate does not reflect the true productive value or opportunity cost of commodities.

I concluded that we need to face up directly to the question of discounting future wellbeing. Some arguments have been offered in favour of a positive rate of discount, and others in favour of a zero rate. I examined the arguments, and, on the basis of present evidence, came down in favour of a zero rate.

Finally, in Chapter 4, I came to the question of how demographic effects should be accommodated in the value function. I examined various different views. Each can be expressed as a condition imposed on the form of the value function. I argued that several conditions common in the literature are definitely incorrect. I also mentioned some that I believe to be correct: principally conditions of separability. However, taken together, these conditions imply that the value function has a very specific form, which has been called 'critical level utilitarianism'. But I find critical level utilitarianism hard to believe. So this chapter raised a difficulty that I left unsolved.

Bibliography

Arrow, Kenneth J., *Social Choice and Individual Values*, Second Edition, Cowles Foundation, 1963.

Arthur, W. B., 'The economics of risk to life', *American Economic Review*, (1981), pp. 54–64.

Atkinson, Anthony B., and Stiglitz, Joseph E., *Lectures on Public Economics*, McGraw–Hill, 1980.

Barry, Brian, 'Circumstances of justice and future generations', in R. I. Sikora and Brian Barry (eds), *Obligations to Future Generations*, Temple University Press, 1978, pp. 204–48.

Barry, Brian, 'Intergenerational justice in energy policy', in Douglas MacLean and Peter G. Brown (eds), *Energy and the Future*, Rowman and Littlefield, 1983, pp. 15–30.

Barry, Brian, 'Rawls on average and total utility: a comment', *Philosophical Studies*, 31 (1977), pp. 317–25.

Barry, Brian, *Theories of Justice*, Harvester–Wheatsheaf, 1989.

Bayles, Michael D., (ed.), *Ethics and Population*, Schenkman, 1976.

Beckerman, Wilfred, 'Global warming: a sceptical economic assessment', in Dieter Helm (ed.), *Economic Policy Towards the Environment*, Blackwell, 1991, pp. 52–85.

Bergstrom, T., Blume, L., and Varian, H., 'On the private provision of public goods', *Journal of Public Economics*, 29 (1986), pp. 25–50.

Berz, G., 'Climatic change: impact on international reinsurance', in G. I. Pearman (ed.), *Greenhouse: Planning for Climate Change*, CSIRO, Australia, 1988, pp. 579–87.

135

Blackorby, Charles, and Donaldson, David, 'Social criteria for evaluating population change', *Journal of Public Economics*, 25 (1984), pp. 13–33.

Broeker, Wallace S., and Denton, George H., 'What drives glacial cycles?', *Scientific American*, (January 1990), pp. 42–50.

Broome, John, 'The economic value of life', *Economica*, 52 (1985), pp. 281–94.

Broome, John, 'Fairness', *Proceedings of the Aristotelian Society*, 91 (1990), pp. 87–102.

Broome, John, 'Some principles of population', in David Collard, David Pearce and David Ulph (eds), *Economics, Growth and Sustainable Environments*, Macmillan, 1988, pp. 85–96.

Broome, John, *Weighing Goods*, Blackwell, 1991.

Budd, W. F., 'The expected sea-level rise from climatic warming in the Antarctic', in G. I. Pearman (ed.), *Greenhouse: Planning for Climate Change*, CSIRO, Australia, 1988, pp. 74–82.

Callendar, G. S., 'The artificial production of carbon dioxide', *Quarterly Journal of the Royal Meteorological Society*, 64 (1938), pp. 223–40.

Cowen, Tyler, 'Normative population theory', *Social Choice and Welfare*, 6 (1989), pp. 33–43.

Cowen, Tyler, 'A positive argument for a zero rate of intergenerational discount', in Peter Laslett and James Fishkin (eds), *Philosophy, Politics and Society: Series VI, Future Generations*, Yale University Press, forthcoming.

Cowen, Tyler, and Parfit, Derek, 'Against the social discount rate', in Peter Laslett and James Fishkin (eds), *Philosophy, Politics and Society: Series VI, Future Generations*, Yale University Press, forthcoming.

Dasgupta, Partha, 'Lives and well-being', *Social Choice and Welfare*, 5 (1988), pp. 103–26.

Dasgupta, Partha, 'Population and savings: normative issues', typescript, 1991.

Dasgupta, Partha, 'Population size and the quality of life', *Aristotelian Society Supplementary Volume*, 63 (1989), pp. 23–54.

Dasgupta, Partha, 'Resource depletion, research and development, and the social rate of discount', in Robert C. Lind et al., *Discounting for Time and Risk in Energy Policy*, Resources for the Future, 1982, pp. 273–305.

Dasgupta, Partha, and Heal, G. M., *Economic Theory and Exhaustible Resources*, Cambridge University Press, 1979.

Department of Transport, *Values for Journey Time Savings and Accident Prevention*, pamphlet, 1987.

Diamond, Peter A., 'The evaluation of infinite utility streams', *Econometrica*, 33 (1965), pp. 170–7.

Diamond, Peter A., and Mirrlees, James A., 'Optimal taxation and public production. I: production efficiency', *American Economic Review*, 61 (1971), pp. 8–27.

Dworkin, Ronald, 'What is equality?', *Philosophy and Public Affairs*, 10 (1981), pp. 185–246 and 283–345.

Eckstein, O., 'Investment criteria for economic development and the theory of intertemporal welfare economics', *Quarterly Journal of Economics*, 71 (1957), pp. 56–85.

Feldstein, M. S., 'The social time preference discount rate in cost-benefit analysis', *Economic Journal*, 74 (1964), pp. 360–9. Reprinted in Richard Layard (ed.), *Cost-Benefit Analysis*, Penguin, 1972, pp. 245–69.

French, S., 'Group consensus using expert opinions', in Bernardo et al., *Bayesian Statistics 2*, 1985, pp. 183–202.

Gauthier, David, *Morals by Agreement*, Oxford University Press, 1985.

Gorman, W. M., 'The structure of utility functions', *Review of Economic Studies*, 35 (1968), pp. 367–90.

Graaf, J. de V., *Theoretical Welfare Economics*, Cambridge University Press, 1957.

Haines, Andrew, 'The implications for health', in Jeremy Leggett (ed.), *Global Warming: The Greenpeace Report*, Oxford University Press, 1990, pp. 149–62.

Hammond, Peter J., 'Consequentialist demographic norms and parenting rights', *Social Choice and Welfare*, 5 (1988), pp. 127–45.

Harsanyi, John C., 'Cardinal welfare, individualistic ethics, and interpersonal comparisons of utility', *Journal of Political Economy*, 63 (1955), pp. 309–21.

Harsanyi, John C., *Rational Behavior and Bargaining Equilibrium in Games and Social Situations*, Cambridge University Press, 1977.

Harvey, Charles, 'Valuing future costs and benefits', typescript, 1992.

Houghton, J. T., Jenkins, G. J., and Ephraums, J. J., (eds), *Climate Change: The IPCC Scientific Assessment*, Cambridge University Press, 1990.

Hume, David, *An Enquiry Concerning the Principles of Morals*,

1751.

Huntley, Brian, 'Lessons from climates of the past', in Jeremy Leggett (ed.), *Global Warming: The Greenpeace Report*, Oxford University Press, 1990, pp. 133–48.

Intergovernmental Panel on Climate Change, *Overview and Conclusions: Climate Change, A Key Global Issue*, World Meteorological Organization and United Nations Environment Programme, Draft, 1990.

Intergovernmental Panel on Climate Change, *Policymakers' Summary: Working Group III*, World Meteorological Organization and United Nations Environment Programme, Draft, 1990.

Intergovernmental Panel on Climate Change, *Scientific Assessment of Climate Change: The Policymakers' Summary*, World Meteorological Organization and United Nations Environment Programme, 1990.

Jamieson, Dale, 'Managing the future: public policy, scientific uncertainty, and global warming', in Donald Sherer (ed.) *Upstream / Downstream: Issues in Environmental Ethics*, Temple University Press, 1991, pp. 67–89.

Jones–Lee, M. W., *The Economics of Safety and Physical Risk*, Blackwell, 1989.

Koopmans, Tjalling C., 'Representation of preference orderings over time', in C. B. McGuire and R. Radner (eds), *Decisions and Organization: A Volume in Honour of Jacob Marschak*, North–Holland, 1972, pp. 79–100.

Laslett, Peter, 'Is there a generational contract?', in Peter Laslett and James Fishkin (eds), *Philosophy, Politics and Society: Series VI, Future Generations*, Yale University Press, forthcoming.

Lecomber, Richard, *The Economics of Natural Resources*, Macmillan, 1979.

Leggett, Jeremy, 'The nature of the greenhouse threat', in Jeremy Leggett (ed.), *Global Warming: The Greenpeace Report*, Oxford University Press, 1990, pp. 14–43.

Lind, Robert C., 'A primer on the major issues relating to the discount rate for evaluating national energy policy', in Robert C. Lind et al., *Discounting for Time and Risk in Energy Policy*, Resources for the Future, 1982, pp. 21–94.

Lind, Robert C., 'Reassessing the government's discount rate policy in the light of new theory and data in a world economy with a high degree of capital mobility', *Journal of Environmental Economics and Management*, 18 (1990), pp. S8–S28.

MacLean, Douglas, 'A moral requirement for energy policy', in

Douglas MacLean and Peter G. Brown (eds), *Energy and the Future*, Rowman and Littlefield, 1983, pp. 180–97.

Marglin, S. A., 'The social rate of discount and the optimal rate of investment', *Quarterly Journal of Economics*, 77 (1963), pp. 95–111.

Marshall, Alfred, *Principles of Economics*, Eighth Edition, Macmillan, 1920.

McKerlie, Dennis, 'Equality and time', *Ethics*, 99 (1989) 475–91.

McMahan, J. A., 'Problems of population theory', *Ethics*, 92 (1981).

Morton, Adam, *Disasters and Dilemmas*, Blackwell, 1991.

Nagel, Thomas, *The Possibility of Altruism*, Oxford University Press, 1970.

Narveson, Jan, 'Moral problems of population', *The Monist*, 57 (1973), reprinted in Michael D. Bayles (ed), *Ethics and Population*, Schenkman, 1976, pp. 59–80.

Narveson, Jan, 'Utilitarianism and new generations', *Mind*, 76 (1967), pp. 62–72.

Newbery, David M., 'The isolation paradox and the discount rate for benefit-cost analysis', *Quarterly Journal of Economics*, 105 (1990), pp. 235–8.

Ng, Yew–Kwang, 'Social criteria for evaluating population change: an alternative to the Blackorby–Donaldson criterion', *Journal of Public Economics*, 29 (1986), pp. 375–81.

Nordhaus, W. D., 'To slow or not to slow: the economics of the greenhouse effect', *Economic Journal*, 101 (1991), pp. 920–37.

Nozick, Robert, *Anarchy, State and Utopia*, Basic Books, 1974.

Olson, Mancur, and Bailey, Martin J., 'Positive time preference', *Journal of Political Economy*, 89 (1981), pp. 1–25.

Page, Talbot, 'Intergenerational equity and the social rate of discount', in V. Kerry Smith (ed.), *Environmental Resources and Applied Welfare Economics: Essays in Honor of John V. Krutilla*, Resources for the Future, 1988, pp. 71–89.

Page, Talbot, 'Intergenerational justice as opportunity', in Douglas MacLean and Peter G. Brown (eds), *Energy and the Future*, Rowman and Littlefield, 1983, pp. 38–58.

Parfit, Derek, *Reasons and Persons*, Oxford University Press, 1984.

Pearce, David, 'Ethics, irreversibility, future generations and the social rate of discount', *International Journal of Environmental Studies*, 21 (1983), pp. 67–86.

Pearce, David, Barbier, Edward, and Markandya, Anil, *Sustainable Development: Economics and Environment in the Third World*, Edward Elgar, 1990.

Pezzey, John, 'Sustainability, intergenerational equity, and environmental policy', typescript, University of Bristol, 1989.

Pigou, A. C., *The Economics of Welfare*, Fourth Edition, Macmillan, 1932.

Rawls, John, *A Theory of Justice*, Oxford University Press, 1972.

Ramsey, Frank, 'A mathematical theory of saving', *Economic Journal*, 38 (1928), pp. 543–9, reprinted in his *Foundations: Essays in Philosophy, Logic, Mathematics and Economics*, edited by D. H. Mellor, Routledge and Kegan Paul, 1978, pp. 261–81.

Regan, Donald H., 'Against evaluator relativity: a response to Sen', *Philosophy and Public Affairs*, 12 (1983), pp. 93–112.

Richards, David A. J., 'Contractarian theory, intergenerational justice, and energy policy', in Douglas MacLean and Peter G. Brown (eds), *Energy and the Future*, Rowman and Littlefield, 1983, pp. 131–50.

Savage, Leonard J., *The Foundations of Statistics*, Second Edition, Dover, 1972.

Scheffler, Samuel, (ed.), *Consequentialism and Its Critics*, Oxford University Press, 1988.

Scheffler, Samuel, *The Rejection of Consequentialism*, Oxford University Press, 1982.

Schelling, Thomas C., 'Climatic change: implications for welfare and policy', in *Changing Climate: Report of the Carbon Dioxide Assessment Committee*, National Academy Press, 1983, pp. 449–82.

Schimel, David, 'Biogeochemical feedbacks in the Earth system', in Jeremy Leggett (ed.), *Global Warming: The Greenpeace Report*, Oxford University Press, 1990, pp. 68–82.

Schneider, Stephen H., 'The changing climate', *Scientific American*, (September 1989), pp. 38–47.

Schneider, Stephen H., 'The science of climate-modelling and a perspective on the global-warming debate', in Jeremy Leggett (ed.) *Global Warming: The Greenpeace Report*, Oxford University Press, 1990, pp. 44–67.

Schwartz, Thomas, 'Obligations to posterity', in R. I. Sikora and Brian Barry (eds), *Obligations to Future Generations*, Temple University Press, 1978, pp. 3–13.

Sen, Amartya K., 'Approaches to the choice of discount rates for social benefit-cost analysis', in Robert C. Lind et al., *Discounting for Time and Risk in Energy Policy*, Resources for the Future, 1982, pp. 325–51.

Sen, Amartya K., 'Introduction', in his *Resources, Values and*

Development, Blackwell, 1984, pp. 1–34.

Sen, Amartya K., 'Isolation, assurance, and the social rate of discount', *Quarterly Journal of Economics*, 81 (1967), pp. 112–24.

Sen, Amartya K., 'Rights and agency', *Philosophy and Public Affairs*, 11 (1982), pp. 3–38.

Sen, Amartya K., 'Utilitarianism and welfarism', *Journal of Philosophy*, 76 (1979), pp. 463–89.

Sidgwick, Henry, *The Methods of Ethics*, Seventh edition, Macmillan, 1907.

Sikora, R. I., 'Is it wrong to prevent the existence of future generations?', in R. I. Sikora and Brian Barry (eds), *Obligations to Future Generations*, Temple University Press, 1978, pp. 112–66.

Slote, Michael, *Beyond Optimizing: A Study of Rational Choice*, Harvard University Press, 1989.

Smith, Adam, *An Inquiry Into the Nature and Causes of the Wealth of Nations*, Everyman Edition, 1910.

Smith, Michael, 'The Humean theory of motivation', *Mind*, 96 (1987), pp. 37–41.

Solow, Robert, 'The economics of resources or the resources of economics', *American Economic Review Papers and Proceedings*, (1974), pp. 1–14.

Spackman, Michael, 'Discount rates and rates of return in the public sector: economic issues', Government Economic Service Working Paper 113.

Stark, K. P., 'Designing for coastal structures in a greenhouse age', in G. I. Pearman (ed.), *Greenhouse: Planning for Climate Change*, CSIRO, Australia, 1988, pp. 161–76.

Stiglitz, Joseph E., 'The rate of discount for benefit-cost analysis and the theory of second best', in Robert C. Lind et al., *Discounting for Time and Risk in Energy Policy*, Resources for the Future, 1982, pp. 151–204.

Strotz, R. H., 'Myopia and inconsistency in dynamic utility maximization', *Review of Economic Studies*, 23 (1955–6), pp. 165–80.

Sumner, L. W., 'Classical utilitarianism and the population optimum', in R. I. Sikora and Brian Barry (eds), *Obligations to Future Generations*, Temple University Press, 1978, pp. 91–111.

Temkin, Larry S., 'Intergenerational inequality', in Peter Laslett and James Fishkin (eds), *Philosophy, Politics and Society: Series VI, Future Generations*, Yale University Press, forthcoming.

Temkin, Larry S., 'Intransitivity and the mere addition paradox', *Philosophy and Public Affairs*, 16 (1987), pp. 138–87.

Tooley, Michael, *Abortion and Infanticide*, Oxford University Press, 1983.

Wagner, Carl G., 'Consensus for belief functions and related uncertainty measures', *Theory and Decision*, 26 (1989), pp. 295–304.

Warr, Peter G., and Wright, Brian D., 'The isolation paradox and the discount rate for benefit-cost analysis', *Quarterly Journal of Economics*, 95 (1981), pp. 129–45.

Wigley, T. M. L., Jones, P. D., and Kelly, P. M., 'Warm world scenarios and the detection of climatic change induced by radiatively active gases', in Bert Bolin, Bo R. Döös, Jill Jäger and Richard A. Warrick (eds), *The Greenhouse Effect, Climatic Change, and Ecosystems*, Wiley, 1986, pp. 271–322.

Woodwell, George M., 'The effects of global warming', in Jeremy Leggett (ed.), *Global Warming: The Greenpeace Report*, Oxford University Press, 1990, pp. 116–32.

Index

143